Stars: Universe's Chemical Factories

Molina

Contents

1. Introduction and motivation

The Universe as we know it contains $\sim 68\%$ dark energy, $\sim 27\%$ dark matter, and $\sim 5\%$ baryonic matter (Planck Collaboration 2014). Even though the fraction seems to be small, all stars and planets are composed of baryonic matter. Studying the composition of stars and the chemical enrichment of our universe is a fascinating and complex research field to answer fundamental questions.

Where do elements come from and where are they formed? The investigation of stars involves interactions on small scales on the level of nucleons to models of whole stellar atmospheres. Therefore, theoretical astrophysics connects the largest topics of modern physics, e.g., gravity, nuclear physics, atomic physics, and optics.

The effort of studying such a broad field has a huge gain, by knowing the composition of individual stars, but also the composition of whole galaxies, one can learn about the astrophysical events that produces all elements that we know. The direct comparison of theoretical models and observations is advantageous as it can indicate whether our understanding of a particular aspect is satisfying or at what point we still lack knowledge. In this way, stars can be used as giant laboratories to probe our knowledge.

While lighter elements up to iron can be produced by many processes, the primary production of heavier elements can raise hints on individual processes. The rapid neutron-capture process (r-process) is responsible for approximately half of all elements heavier than iron in our sun (e.g., Thielemann et al. 2017, Frebel 2018, Cowan et al. 2019, Arnould and Goriely 2020, for recent reviews of the r-process). Some elements as, e.g., gold are made almost exclusively by this process (95%, Sneden et al. 2008). The theoretical foundation of the r-process was proposed already by Burbidge et al. (1957) and Cameron (1957), but its astrophysical production site is still a matter of debate.

Collecting information about the astrophysical production site is a rigorous and interdisciplinary work. The chemical enrichment of a large sample of stars allows to investigate the time dependence of the host event of the r-process. How early is the onset of the production of heavy elements? To answer this question, the chemical history of the Milky Way can provide important clues. In addition to the massive Milky Way, dwarf galaxies are important. These small systems develop with their own chemical history, which makes it possible to investigate early fingerprints of the r-process in the chemical enrichment in many systems. How frequent does such an event appear? A large scatter in the enrichment of heavy elements in old stars as well as the discovery of the highly r-process enriched dwarf galaxies Reticulum II (Ji et al. 2016a, Roederer et al. 2016) and possibly Tucana III (Hansen et al. 2017) provided evidence for a rare production site.

Many events have been suggested as possible hosts for the r-process, the most promising being merging neutron stars (see, e.g., Cowan et al. 2019, Shibata and Hotokezaka 2019, for recent reviews), collapsars (e.g., Nakamura et al. 2013), and magneto-rotational driven supernovae (e.g., Thielemann et al. 2017). Neutron star mergers (NSMs) were recently confirmed as host for the r-process. This was possible by the gravitational wave detection GW170817 (LIGO Scientific Collaboration and Virgo Collaboration

2017) and the detected optical counterpart AT2017gfo (e.g., Cowperthwaite 2017, Kasliwal et al. 2017, Chornock et al. 2017, Drout et al. 2017, Shappee et al. 2017, Pian et al. 2017, Waxman et al. 2018, Watson et al. 2019), which showed fingerprints of the r-process. However, the question whether NSMs are also the dominant production source of heavy elements in the early Universe remains open (e.g., Beniamini et al. 2016, 2018, Duggan et al. 2018, Côté et al. 2019a, Skúladóttir et al. 2019, Reichert et al. 2020).

We tackle this question from two perspectives, on the one hand from an observational point of view and on the other hand from a nucleosynthesis perspective. Therefore, we give an overview over the basic astrophysical background in chapter 2, the theoretical foundation of the synthesis of elements in chapter 3, and the foundation to analyze stellar spectra in chapter 4. A homogenized abundance analysis of 13 dwarf galaxies is presented in chapter 5. Afterwards in chapter 6, we discuss abundance trends of the Milky Way and the implications for a possible host of the r-process with the help of Galactic chemical evolution models. To see whether magneto-rotational driven supernovae are able to host the r-process from a theoretical point of view, we present a nucleosynthesis analysis of such an event in chapter 7. Afterwards in chapter 8, we investigate a star with a peculiar chemical composition and try to explain its formation by a superposition of many processes. We end with a summary and conclusion in chapter 9.

2. Astrophysical background

2.1. The origin of elements

After the big bang, the universe consists of the lightest elements, hydrogen (^{1}H $\sim$ 75%) and helium (^{4}He $\sim$ 25%), to smaller extents also of lithium and beryllium (see, e.g., Beringer 2012, Cyburt et al. 2016, for reviews of big bang nucleosynthesis). These nuclei are distributed in clouds of gas that clumped together under the influence of gravity and formed the first stars that are usually referred to as population III stars (Pop III). These stars could possibly become very massive ($\gtrsim 100\,M_\odot$, see e.g., Hartwig et al. 2018, Clark et al. 2011, for a discussion of stellar masses) and they hydrostatically burn light nuclei to heavier nuclei. At the end of their lifetime, they die in a supernova (e.g., Fraley 1968, Chen 2015, Belkus et al. 2007). Within these violent events, matter consisting of all previously synthesized elements up to the iron group is ejected into space again. This matter is present as dust and serves as a nursery of new stars.

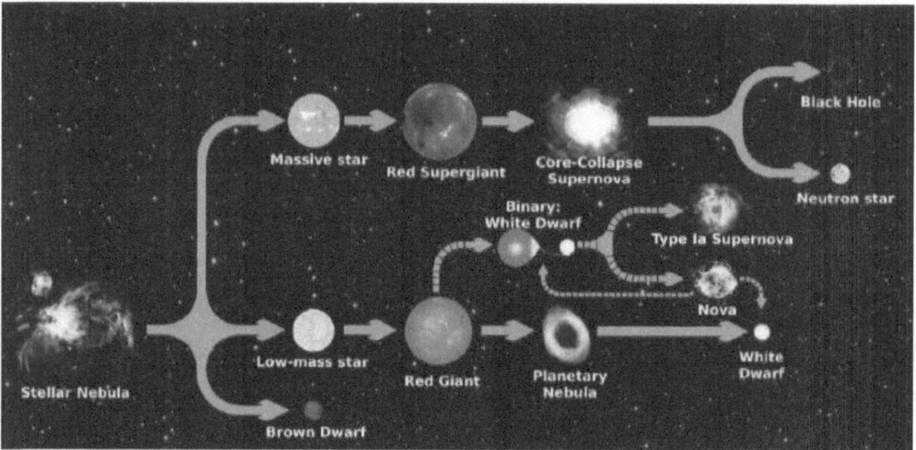

Figure 2.1.: Life cycle of stars. Stars are formed out of dust from previous generations of stars. The final fate strongly depends on their mass. Adapted from Bailey (2020) and courtesy of R.N. Bailey.

The stages of the life cycle of stars are the formation of stars from gas clouds, the synthesis of elements in these stars, and the ejection of this material either as mass loss during stellar evolution or at the end of their lifetimes. Therefore, with increasing time, new generation of stars contain more metals than older generation of stars. In galaxies that still form stars as our Milky Way, this cycle repeats until today (a schematic overview of one iteration is shown in Fig. 2.1). In contrast to the first stars, the following generations of stars are less massive and the final fate of them is mostly determined by the mass of the star. Light stars will form a white dwarf (WD), whereas more massive stars will end their life in a core-collapse supernova (CC-SN) and possibly form a black hole (BH) or a neutron star (NS).

In the Milky Way, a large fraction of stars are thought to be born in a binary system (Heintz 1969, Abt and Levy 1976, Abt 1983, Duquennoy and Mayor 1991, Gao et al. 2014). A binary system allows for the occurrence of additional events such as type Ia supernovae, novae, neutron star mergers, or BH-NS mergers. Every event produces unique chemical fingerprints that are mixed into the interstellar medium. The typical features of each event will be described in the next sections.

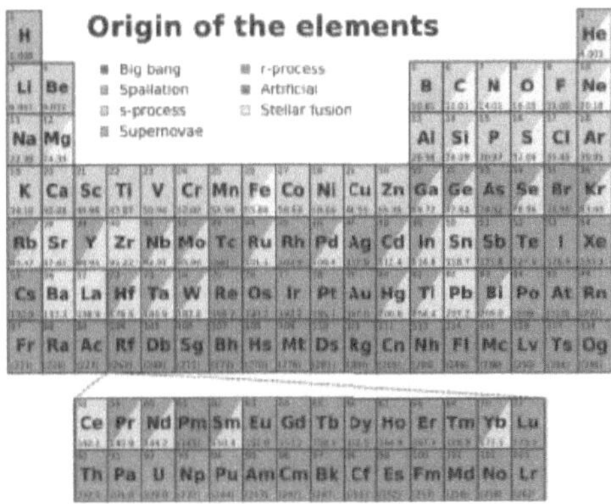

Figure 2.2.: Periodic table. Different colors indicate the dominant origin in the sun according to different production sites and processes. The contributions to heavy elements were taken from (Sneden et al. 2008). Elements with multiple contributions that are larger than 25% are indicated by several colors.

A snapshot of the chemical history of our universe is locked in the photosphere of stars. The composition of e.g., our sun, is a product of many life cycles of stars and events that occurred during the history of our universe until the sun was born. The spectrum of the sun carries dark features (absorption lines) at characteristic wavelengths, caused by the transition of electrons between (element specific) energy levels of the atom (Fraunhofer 1817). The shape and depth of these features gives insight into the composition of the photosphere. A comparison of the solar composition with models of nucleosynthesis processes is useful to determine the dominant astrophysical contribution of each individual element (Fig. 2.2). For the heavy elements (with atomic number $Z \geq 31$) one separates the synthesis of elements in two dominant processes[1], the s- and the r-process (see Sect. 2.3), which are thought to occur in low-mass stars and neutron star mergers, respectively.

---

[1]Note that this is only a simplified view on the individual production sites of heavy elements as the lighter heavy elements ($31 \leq Z \leq 47$) may form also by additional processes (see e.g., Fröhlich et al. 2006, Sneden et al. 2008, Arcones and Montes 2011, Hansen et al. 2014b).

The details of the nucleosynthesis and astrophysical sites that dominantly contributed to the chemical enrichment of our sun will be described in the following sections.

2.2. Stellar evolution and the synthesis of elements up to the iron group

Nucleosynthesis up to the iron group appears quite frequently in our Universe. Under the gravity of a star, temperatures and densities are sufficient for light elements to fuse into heavier ones. However, for heavier nuclei (with larger amount of protons) charged particle reactions become more and more unlikely as the Coulomb barrier increases. In addition, fusion of these elements by charged particle reactions will consume rather than release energy for most reactions[2], because the binding energy per nucleon has its maximum at the iron group (Fig. 2.3).

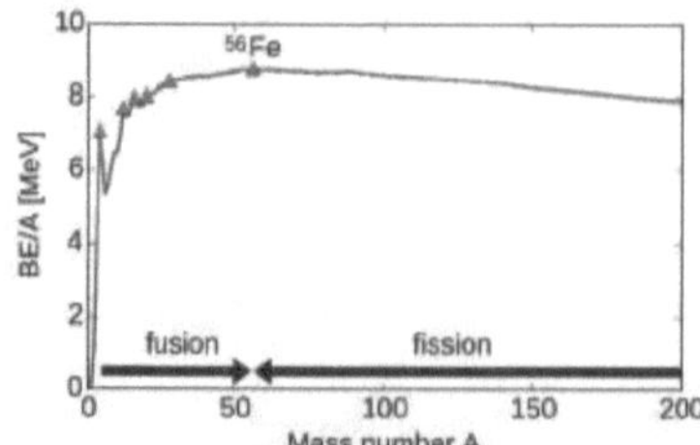

Figure 2.3.: Binding energy per nucleon (BE/A) versus mass number. Energy will be released by fusion of elements lighter than the iron group and by fission of heaver nuclei.

Therefore, during the lifetime of massive stars, iron is one of the heaviest element that is synthesized. All other processes rely on a neutron source in order to synthesize heavier elements (see Sect. 2.3). Figure 2.4 shows different nuclear processes and their typical path in the nuclear chart. In the following, we will describe the different burning stages during the lifetime of a star. The nucleosynthesis of these hydrostatic burning phases is schematically shown as orange arrow in Fig. 2.4.

---

[2]Some charged particle reactions may still provide energy also for heavier nuclei as α-captures

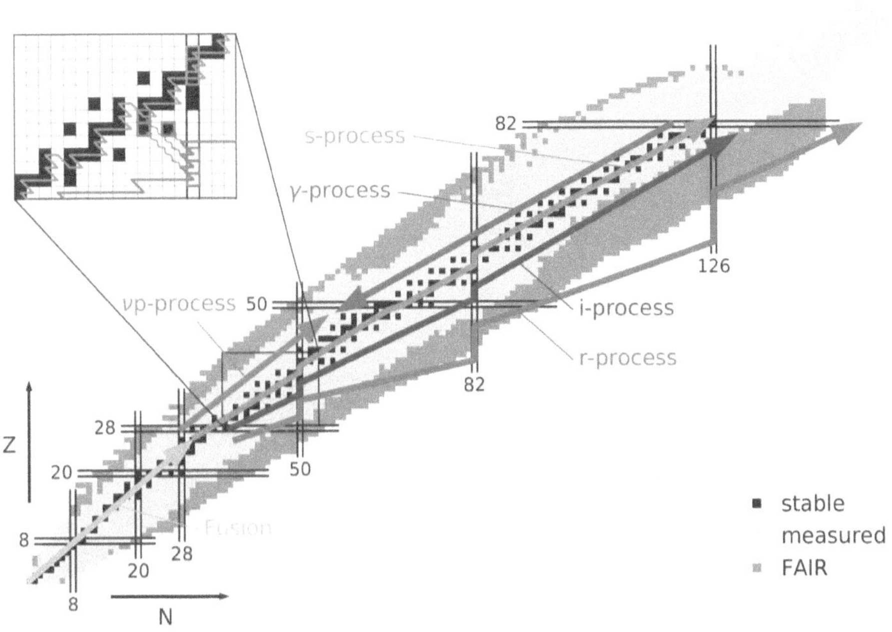

Figure 2.4.: Path of nuclear fusion, s-process, i-process, νp-process, γ-process, and r-process in the nuclear chart. Shown is the neutron number versus the proton number. Courtesy of M. Jacobi.

2.2.1. The evolution of stars

Inside a star, the counterpart to the gravitational pressure is the pressure generated by nuclear fusion as well as degeneracy pressure. This prevents stars from a gravitational collapse. During its life, a star will undergo different burning stages (see e.g., Weaver et al. 1978, Woosley and Weaver 1995, Kippenhahn et al. 2012). First, stars fuse hydrogen into helium. When hydrogen is depleted to a point where its burning does not provide enough energy anymore to act against the gravitational force, the star contracts. This contraction continues until the core reaches a critical central temperature ($\sim 10^8$ K, depending on the mass of the star) to ignite helium. Then, carbon and oxygen get synthesized in the core of the star while hydrogen burning occurs in an outer shell. This provides enough energy to let the star expand again. If the star is massive enough ($\gtrsim 8\,M_\odot$), the process of lapsed burning, contraction, and ignition of the ashes from the previous burning stage is repeated several times. Massive stars are able to undergo hydrogen, helium, carbon, neon, oxygen, and silicon burning, while lighter stars (as e.g., our sun) may end at an earlier burning stage due to lower central temperatures (Fig. 2.5). These lighter stars are finally stabilized by the degeneracy pressure of electrons after shedding their outer shells (planetary nebulae). Astrophysical objects of this type are called white dwarfs (WD).

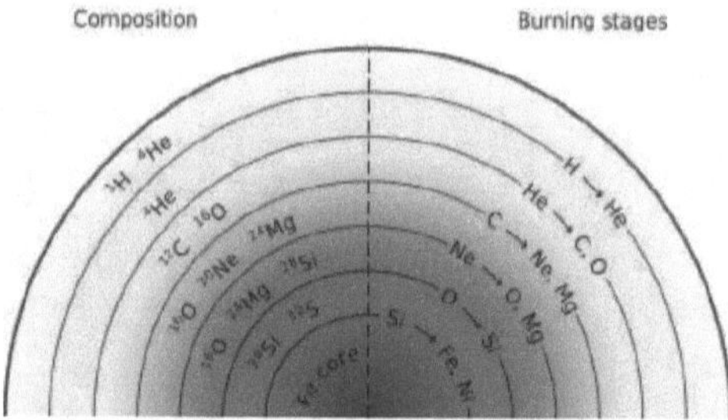

Figure 2.5.: Composition (left) and burning stages (right) in a massive star (Martin 2017).

The burning phases have an influence of the radius and magnitude of the star. Later burning stages where more energy is released lead to a star with a large radius and therefore low surface gravity. Stars of these type are called "Giants".

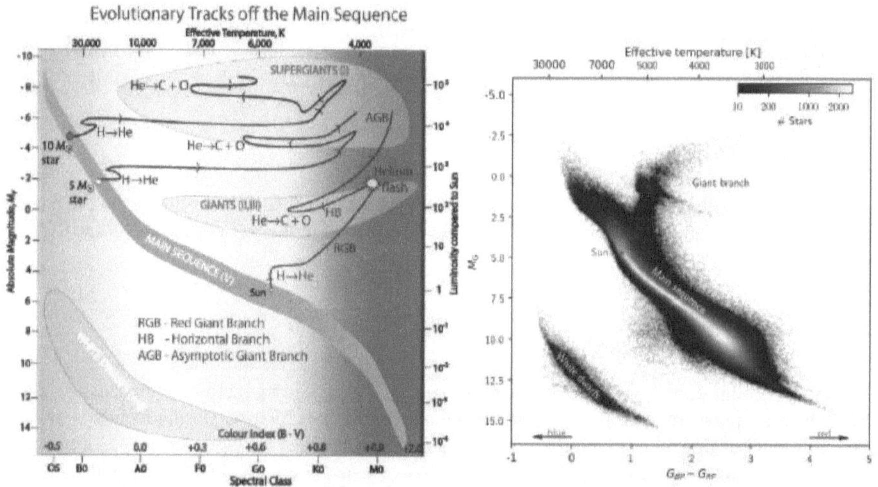

Figure 2.6.: Left Panel: Schematic HR diagram together with evolutionary tracks off the main sequence (taken from Australia Telescope National Facility 2020). Right panel: HR diagram for stars measured by *Gaia* within 500 pc. Estimated values for the sun were taken from Casagrande and VandenBerg (2018) as it is to bright to get measured directly. Stars with uncertain magnitudes, parallaxes, and extinctions as given in (Gaia Collaboration 2018a) are neglected in the plot.

Therefore, observationally, the burning stage of a star can be identified by its color and magnitude´ (Searle et al. 1973). With time, stars will evolve along paths in the color-magnitude plane. The path is determined by the mass of the star (left panel of Fig. 2.6). These types of plots are called Hertzsprung-Russel (HR) diagram (Hertzsprung 1976, Russell 1914) and are also particular useful to determine whether an investigated star is a dwarf (a compact star) or a giant (a spatially extended star). Hydrogen burning is the longest phase in the life of a star. Therefore this phase is called main sequence and most stars are distributed along the main sequence as shown in the right panel of Fig. 2.6, which shows stars measured with *Gaia* (Gaia Collaboration 2018b) within $500\,pc$ neglecting stars with uncertain measurements as given in Gaia Collaboration (2018a). Giant stars will be located in the upper right corner of the diagram which corresponds to red colors, low effective temperatures, and bright stars, whereas the remnants of lighter stars are located in the lower left corner of the HR-diagram that corresponds to blue, high effective temperature, and faint objects.

2.3. The synthesis of elements beyond the iron group

The dominant processes that contribute to the enrichment of heavy elements are the rapid neutron-capture process (r-process) and the slow neutron-capture process (s-process). Indicated already by their name, these processes are characterized by the timescale of neutron-captures and β-decays.

If neutron-captures occur much faster than β-decays, the nucleosynthetic path proceeds along the neutron-rich side (red arrow in Fig. 2.4). This process can only be hosted in extreme environments, where many free neutrons are present. The final nucleosynthesis pattern that results from a r-process will include peaks that are caused by closed neutron shells, the so-called magic numbers. Due to the higher neutron separation energy at these neutron numbers matter more likely β-decays instead of capture further neutrons. Closer to stability, β-decays occur slower and therefore matter will pile up. When it finally decays to stability it forms three abundance peaks. The first r-process peak arises from the magic number at $N = 50$ (see upper left panel of Fig. 2.4), the second r-process peak from $N = 82$, and the third peak from $N = 126$.

Finding hints on the astrophysical hosts of the r-process is a major effort and will be addressed within this work (for reviews of the r-process and the production of heavy elements see e.g., Cowan et al. 1991, Sneden et al. 2008, Thielemann et al. 2011, 2017, Frebel 2018, Horowitz et al. 2019, Cowan et al. 2019, Kajino et al. 2019, Arnould and Goriely 2020). For this, we call two promising candidates to mind, magneto-rotationally driven supernovae (Sect. 2.3.1) and neutron star merger (Sect. 2.3.2).

On the other hand, the s-process takes place over a much longer timescale. During the evolution of low-mass stars, a small amount of neutrons is released via (α,n)-reactions on ^{13}C and ^{22}Ne. These neutrons are captured on previously formed seed nuclei (e.g., iron). With relatively low neutron densities, the amount of neutron-captures occurs much slower than a β-decay (for reviews see e.g., Busso et al. 1999, Käppeler et al. 2011). The nucleosynthetic path therefore proceeds along the valley of stability (blue arrow in Fig. 2.4) and the s-process peaks are located towards higher mass numbers compared to the r-process (see upper left panel of Fig. 2.4).

However, both processes alone cannot describe the presence of all heavy nuclei today, especially proton-rich nuclei. Therefore, other process as e.g., the γ- or p-process (see e.g., Howard et al. 1991, Goriely et al. 2002, Pignatari et al. 2016), and the νp-process (e.g., Fröhlich et al. 2006, Pruet et al. 2006, Wanajo 2006), have been proposed (Fig. 2.4). To explain unusual abundance pattern of individual stars, also the intermediate neutron-capture process (i-process) has been suggested to play an important role in the chemical enrichment of individual stars (e.g., Cowan and Rose 1977, Dardelet et al. 2015, Hampel et al. 2016, Denissenkov et al. 2017, 2019, Côté et al. 2018a, Herwig et al. 2011).

2.3.1. Core-collapse supernovae

The end of the life of a massive star ($\sim 8\,M_\odot$) is marked by an explosive event (for reviews see e.g., Burrows 2000, Woosley and Janka 2005, Janka et al. 2007, Janka 2012, Burrows 2013). Once the silicon in the core of the star has been burned to iron, thermonuclear reactions cease in the center and the gravitational pressure of the star can no longer be counteracted. This situation is schematically shown in the upper left panel of Fig. 2.7. The stabilizing mechanism of an iron core of a star is the pressure of a degenerate electron gas. This is similar to a WD, whose pressure is also dominated by degenerated electrons. We can therefore use the mass limit that was discovered by Chandrasekhar (1931) to estimate the critical mass of the iron core above which the collapse occurs. For a non-rotating spherical star it is given by (e.g., Nomoto and Hashimoto 1988, Bethe 1990, Woosley and Weaver 1995, Baron and Cooperstein 1990):

$$M_{\mathrm{Ch}} = (2Y_e)^2 \cdot 1.45727 \cdot \left(1 + \left(\frac{S_e}{\pi Y_e}\right)^2\right) \, M_\odot, \tag{2.1}$$

with the electron fraction Y_e and the electron entropy per baryon S_e. The electron fraction is equivalent to the proton fraction when assuming charge neutrality. The second term in brackets is a correction to account for finite temperatures. After reaching this mass threshold, the star becomes unstable and collapses with a (free fall) timescale of the order of (e.g., Müller et al. 2016)

$$t_{\mathrm{ff}} \approx \sqrt{\frac{\pi}{4G\rho}} \sim 1\,\mathrm{s}. \tag{2.2}$$

The average density at a given mass shell ρ is defined as:

$$\rho = \frac{4\pi M}{3r^3}, \tag{2.3}$$

with the enclosed mass M located at an initial radius r. During the collapse, the central temperature and density increases. Furthermore, the electron capture rates on nuclei and free protons increase and therefore the electron fraction Y_e decreases, reducing thus the pressure of the degenerated electron gas. The collapse continues and for high densities neutrino-electron, neutrino-nucleon, and scattering become important. Therefore, neutrino opacities increase. Densities exceeding $\sim 10^{12}\,\mathrm{g\,cm^{-3}}$ involve a plasma that is opaque for neutrinos (e.g., Bethe 1990). The diffusion time of neutrinos becomes longer than the free fall timescale, neutrinos are trapped, and as a consequence a weak equilibrium establishes. The contraction continues until the core reaches densities of the order of nuclear densities ($\rho_{\mathrm{nuc}} \approx 3 \cdot 10^{14}\,\mathrm{g\,cm^{-3}}$). At such high densities, nuclear interactions involve a repulsive force (e.g., Reid 1968), the core does not contract further and a shock is launched as shown in the upper right corner of Fig. 2.7. This bounce is often taken as a reference point in time and the time of later events is referred to as "time after bounce".

Initially, it was believed that the shock wave carries enough energy to cause an successful explosion without stalling (so-called prompt explosion mechanism, e.g., Hillebrandt et al. 1976). The shock, however, loses energy by photo-dissociation reactions and neutrino emission. Due to the loss of energy, the shock stalls at some point and the mechanism to revive it is still an outstanding problem (e.g., Janka

2012, Burrows 2013, Bethe and Wilson 1985, Janka et al. 2007). Nowadays more explosion mechanisms have been proposed (see Janka 2012, Kotake 2013, for reviews on explosion mechanisms).

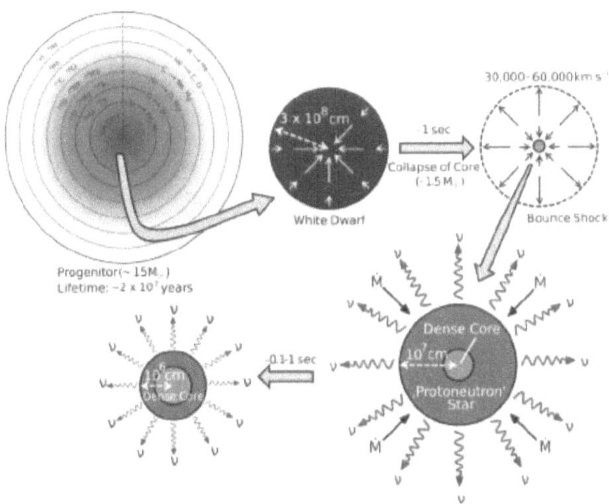

Figure 2.7.: Schematic mechanism of core-collapse supernovae. Shown is the evolution from the progenitor to the neutron star. The iron core of the massive star collapses to a protoneutron star when it reaches the Chandrasekhar mass limit. It finally ends up in a more compact object, a neutron star. Adapted from Burrows (2013).

Another possible explanation for reviving the shock was already proposed by Colgate and White (1966) and Arnett (1966). In this delayed-neutrino heating scenario, neutrinos from the very dense and hot (freshly formed) proto-neutron star (PNS) deposit energy at the shock surface (bottom right of Fig. 2.7, first described in Bethe and Wilson 1985). When the shock revives successfully, it further moves through the layers of the star (Fig. 2.5), heats and ejects them. In these layers, explosive nucleosynthesis occurs. Dominant reactions of this explosive nucleosynthesis are α-captures as the matter in the layers is fairly symmetric (i.e., $Y_e \sim 0.5$). In this way, elements up to the iron group are synthesized. This includes radioactive elements as ^{44}Ti and ^{56}Ni that power the light curve and also leave late-time fingerprints in the X-rays of CC-SNe (e.g., Matz et al. 1988). Caused by the initially high velocities and opacities of the ejecta of CC-SNe, the X-ray signature of the decay of ^{44}Ti is smeared out by scattering processes and the doppler effect. However, since ^{44}Ti is a relatively long lived isotope with a half-life of $\sim 60\,\mathrm{yr}$, the decay of this isotope can be directly observed and the peaks in the spectra resolved after sufficient time (Grebenev et al. 2012, Grefenstette et al. 2014).

The hot PNS in the inner core of the former star cools down by neutrino emission and forms a more compact object. This object is either a black hole (BH) or a neutron star (NS), depending on the structure of the progenitor (e.g., Ugliano et al. 2012, Ertl et al. 2016, Sukhbold et al. 2016).

Magneto-rotational driven Supernovae

The magneto-rotational (MR) or magneto-hydrodynamical driven (MHD) supernovae belong to the class of CC-SNe, but rely on a different explosion mechanism than a "typical" neutrino-driven CC-SNe (a detailed description is given in, e.g., Käppeli 2013).

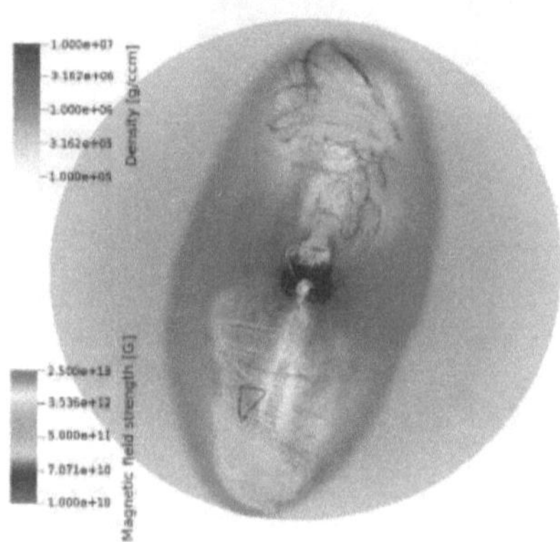

Figure 2.8.: A 3D-simulation (35OC-RO) of an exploding star. Red colors indicate the density, whereas spectral colors indicate the magnetic field strength. Courtesy of M. Obergaulinger.

In these type of events, the presence of a magnetic field involves additional forces that accelerate matter outwards. The magnetic field is amplified by four main mechanisms (even though other amplification effects exist):

(1) By assuming magnetic flux conservation one can derive that the magnetic field in a star amplifies proportional to the density, $\rho^{2/3}$, by compression of the matter alone (e.g., Käppeli 2013).

(2) Due to differential rotation the poloidal component of the magnetic field is wound up around the poles and into the toroidal component. Winding up the magnetic field will consume rotational kinetic energy (e.g., Wheeler et al. 2000, 2002, Käppeli 2013), but amplifies magnetic pressure, in particular along the rotational axis. This winding is clearly visible in Fig. 2.8, which shows the density and magnetic field strength in a recent 3D MR-driven CC-SN simulation.

(3) Another possibility to amplify the magnetic field strength is the magneto-rotational instability (MRI, Balbus and Hawley 1991, 1998, Akiyama et al. 2003, Pessah and Psaltis 2005, Obergaulinger et al. 2009). This instability acts on small spatial scales, but possibly amplifies the magnetic field to dynamically relevant field strengths. In current hydrodynamical simulations of CC-SN, it is difficult to resolve.

(4) The magnetic field can be amplified by convection and the stationary accretion shock instability (SASI, Thompson and Duncan 1993, Endeve et al. 2010, Raynaud et al. 2020).

From a nucleosynthetic point of view, this scenario is quite interesting as it has been suggested to host the r-process (e.g., Meier et al. 1976, Meyer 1994, Nishimura et al. 2006, Winteler et al. 2012, Saruwatari et al. 2013, Nishimura et al. 2015, 2017, Mösta et al. 2018, Halevi and Mösta 2018). In contrast to "typical" neutrino-driven CC-SNe, where matter is processed by neutrinos and therefore neutrons can react to protons (depending on the neutrino properties, i.e., energies and luminosities, and the abundances of neutrons and protons) by:

$$\nu_e + n \rightleftharpoons e^- + p \tag{2.4}$$

$$\bar{\nu}_e + p \rightleftharpoons e^+ + n. \tag{2.5}$$

MR-SNe may eject matter that is dominantly driven by magnetic pressure and therefore conserve neutron rich conditions (see e.g., Nishimura et al. 2015, Mösta et al. 2018). However the existence of such events is not clear as the models require a fast rotation. These rotation rates are not obtained in current progenitor models (Heger et al. 2005). Nevertheless, stars with low metallicity (and therefore low rates of mass-loss and loss of angular momentum) may be able to achieve the necessary rotational velocities (Maeder et al. 1999, Maeder and Meynet 2001, Martayan et al. 2007).

Woosley and Heger (2006) speculate that around 1% of all progenitor stars may host favorable conditions to enable a MR-SNe. One evidence that MR-SNe exist, is the observation of pulsars (see Harding and Lai 2006, for a review) and long γ-ray bursts (Woosley and Bloom 2006). We analyze the nucleosynthesis of these events with state of the art simulations in chapter 7.

2.3.2. Neutron star mergers

A confirmed site of the r-process are neutron star mergers. These systems consist of two neutron stars, previously formed by CC-SNe, that undergo an inspiral phase and finally collide (for recent reviews see e.g., Thielemann et al. 2017, Baiotti and Rezzolla 2017, Cowan et al. 2019, Shibata and Hotokezaka 2019).

During merger, matter is ejected by tidal forces (so-called dynamic ejecta). Consisting mainly out of material from the neutron star itself and being located far away from the colliding neutron stars, the tidally ejected and unshocked matter is to a minimum processed by neutrinos. Therefore it is very neutron rich and an ideal host of the r-process (e.g., Freiburghaus et al. 1999, Rosswog et al. 1999, Goriely et al. 2011, Roberts et al. 2011, Korobkin et al. 2012, Bauswein et al. 2013, Wanajo et al. 2014, Just et al. 2015, Goriely 2015, Bovard et al. 2017, Radice et al. 2018). Closer to the colliding neutron stars, material experiences shocks, resulting in less neutron-rich material. The ejected material may contribute to all r-process elements, including elements of the first r-process peak.

Around the freshly formed hot hyper-massive proto-neutron star (HMNS) or black-hole, an accretion disk forms. Neutrinos that originate in the HMNS and the accretion disk can push matter from the disk outwards in a so-called neutrino-driven wind. This material is less neutron-rich and therefore synthesizes mainly lighter elements between the first and second r-process peak (e.g., Ruffert et al. 1997, Rosswog and Ramirez-Ruiz 2002, Dessart et al. 2009, Wanajo and Janka 2012, Perego et al. 2014, Just et al. 2015, Martin et al. 2015, Lippuner et al. 2017, Fujibayashi et al. 2017).

In the accretion disk, matter can get ejected by viscous heating and due to recombination energy, i.e. energy released by nuclear reactions in the disk when free nucleons combine into helium. (e.g., Metzger et al. 2008, Lee et al. 2009, Fernández and Metzger 2013, Fernández et al. 2015a,b, Just et al. 2015, Wu et al. 2016, Shibata et al. 2017, Lippuner et al. 2017). This ejecta component is very diverse in neutron-richness. Therefore, it may form elements up to the third r-process peak (see e.g., Wu et al. 2016), but also lighter elements like iron. It is thought that 5% to 20% of the initial disk mass is ejected as viscous and neutrino-driven ejecta,(Fernández and Metzger 2013, Fernández et al. 2015a, Just et al. 2015, Martin et al. 2015), where the numbers depend on the final fate of the remnant. If a black-hole forms rapidly, these numbers are smaller compared to a long living HMNS.

Neutron star merger simulations today mostly take neutrinos into account only in an approximate way by using a simplified leakage scheme (This is caused by difficulties with the commonly used closure relation of more detailed neutrino transport schemes as M1. See Foucart et al. 2018, for a review of radiation transport in NSM simulations). However, neutrinos are an important ingredient to calculate the neutron richness and therefore the nucleosynthesis of the ejecta (e.g., Martin et al. 2018). Therefore, all above statements summarize the current knowledge, but our understanding may change in the future. The presence of several nucleosynthetic components, including elements up to the third r-process peak, was demonstrated by several models in order to match the observation of a binary neutron star merger (LIGO Scientific Collaboration and Virgo Collaboration 2017) accompanied by an electromagnetic transient (e.g., Cowperthwaite 2017, Chornock et al. 2017, Kasliwal et al. 2017, Drout et al. 2017, Shappee et al. 2017, Pian et al. 2017, Waxman et al. 2018, Arcavi 2018, Watson et al. 2019) and the change of color from blue to red caused by the high opacities of lanthanides (see Sect. 4.1.3).

Even though observed, there arise several problems when taking neutron star mergers as dominant production site of r-process elements into account. These discrepancies are present when observing the chemical enrichment of stars and are discussed in detail in chapters 5 and 6.

2.4. Observation of stars

When a star forms, it locks a snapshot of the surrounding interstellar medium (ISM) into its photosphere. The abundances in the photosphere are observable via spectra analysis. Therefore, stars can yield insights into the chemical enrichment of the galaxy at their formation time and therefore provide information on the production sites of heavy elements. As the abundance of an element depends on the full set of elements, it can only be expressed in relative units to another reference element or by making other assumptions. The abundance of element E in the photosphere of a star is therefore often given as absolute abundance

$$\log_{10} \epsilon(X) \equiv \log_{10}\left(\frac{A_\mathrm{E}}{A_\mathrm{H}}\right) + 12, \tag{2.6}$$

where A_H is the abundance of hydrogen and is assumed to be 10^{12} per definition. To have a direct comparison to the solar system, the definition of square brackets for abundances of element E and Z is given as

$$[\mathrm{E/Z}] \equiv \log_{10}\left(\frac{A_\mathrm{E}}{A_\mathrm{Z}}\right) - \log_{10}\left(\frac{A_\mathrm{E}}{A_\mathrm{Z}}\right)_\odot, \tag{2.7}$$

where the last term is the elemental fraction in our sun. A substitute for the age of a star is given by [Fe/H], and it is referred to as metallicity. Over time, the Galaxy is enriched with metals and thus the photosphere of the stars. The choice of iron as a proxy for the time is historical, but it is also a good observable as it shows strong absorption features in the stellar spectrum. Large negative metallicities denote old stars, whereas stars with positive metallicities are younger than our sun. Neither square brackets nor absolute abundances have any unit, but sometimes the unit dex clarifies that it is a logarithmic quantity where one dex is one order of magnitude.

2.4.1. Abundances in old stars

Old stars are particularly interesting as their composition originates from only a few or even individual astrophysical events. Figure 2.9 shows the abundance of six r-process enhanced metal-poor stars ([Fe/H] < -2). The abundance ratio of heavy neutron-capture elements ($Z \geq 56$) is fairly constant, whereas the ratio for the lighter heavy elements ($31 \leq Z \leq 50$) scatters (middle and bottom panel in Fig. 2.9). The responsible nucleosynthesis processes should be primary processes as these old stars do not contain many seed nuclei. The s-process does not act at these early times (with the possible exception of some rare events as e.g., rapidly rotating spin stars, Pignatari et al. 2008, Frischknecht et al. 2012, Chiappini 2013), but the r-process can operate early on (Sect. 2.3). The robust elemental ratio of the heavy neutron-capture elements led to the conclusion that only one process with robust elemental ratios contributes to these elements (Sneden et al. 2003).

Abbreviation	Definition
r-I	$0.3 \leq [\mathrm{Eu/Fe}] \leq 1$ and $[\mathrm{Ba/Eu}] < 0.0$
r-II	$[\mathrm{Eu/Fe}] > 1$ and $[\mathrm{Ba/Eu}] < 0.0$
s	$[\mathrm{Ba/Fe}] > 1$ and $[\mathrm{Ba/Eu}] > 0.5$
r/s	$0 < [\mathrm{Ba/Eu}] < 0.5$

Table 2.1.: Definition of subclasses of metal-poor stars based on their neutron-capture elements (Beers and Christlieb 2005).

On the other hand, the lighter heavy elements may need an additional astrophysical source to explain the abundance scatter (middle panel of Fig. 2.9).

Observations of stars in isolated dwarf galaxies show that the enrichment of r-process elements is not uniform. Some dwarf galaxies are highly r-process enriched (e.g., Reticulum II, Ji et al. 2016a and possibly Tucana III, Hansen et al. 2017), whereas others are deficient in neutron-capture elements (e.g., Hercules, Koch et al. 2008 and Triangulum II, Ji et al. 2019).

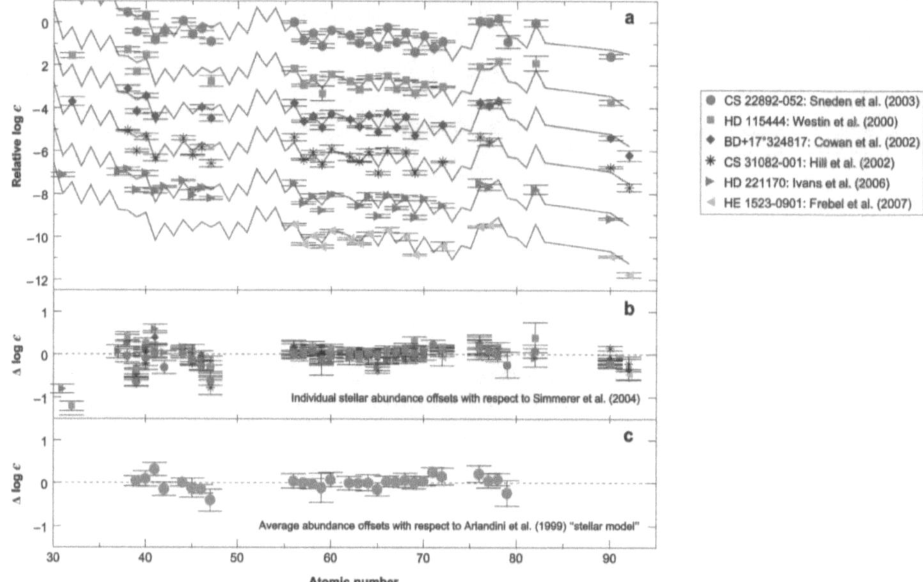

Figure 2.9.: Upper panel (a): Neutron-capture abundances of six r-process enhanced metal-poor stars. The solid light blue lines are the scaled r-only Solar-system elemental abundance curves (Simmerer et al. 2004, Cowan and Sneden 2006). Middle panel (b): Residual abundance offsets with respect to Simmerer et al. (2004). Lower panel (c): Average residual abundance offsets with respect to Arlandini et al. (1999). Adopted from Sneden et al. (2008).

A larger scatter of the [Eu/Fe] ratio of metal-poor stars shows that individual stars are enriched by a large fraction of europium, whereas others do not experience such a strong enrichment. These two observations indicate that r-process elements are produced in rare events (relative to the more frequent events as CC- or type Ia SNe). Stars that are r-process enriched or polluted by the s-process are classified as shown in Tab. 2.1.

Carbon enhanced stars

The lowest ever constrained metallicities in stars are given by $[Fe/H] \leq -7.1$ and $[Fe/H] = -6.2 \pm 0.2$ for SMSS J031300.36-670839.3 (Keller et al. 2014) and SMSS J160540.18-144323 (Nordlander et al. 2019), respectively. These and other extremely metal-poor stars are thought to have formed out of the material from the first generation of stars (Pop III). Similar to the aforementioned stars, most of the extremely metal poor stars show an enhancement in carbon. Stars with low metallicities and peculiar enhancements in carbon are categorized as carbon-enhanced metal-poor stars (CEMP, see Tab. 2.2).

Abbreviation	Definition
CEMP	$[C/Fe] > 1.0$
CEMP-r	$[C/Fe] > 1.0$ and $[Eu/Fe] > 1.0$
CEMP-s	$[C/Fe] > 1.0$, $[Ba/Fe] > 1.0$, and $[Ba/Eu] > 0.5$
CEMP-r/s	$[C/Fe] > 1.0$ and $0.0 < [Ba/Eu] < 0.5$
CEMP-no	$[C/Fe] > 1.0$ and $[Ba/Fe] < 0.0$

Table 2.2.: Definition of Carbon enhanced metal-poor stars based on their neutron-capture elements (Beers and Christlieb 2005).

The origin of CEMP stars is still a matter of debate. A large fraction of binary detections (as visible in the radial velocity profile of a star) lead to the hypothesis that CEMP-s stars accrete matter from an AGB companion star (McClure and Woodsworth 1990, Lucatello et al. 2005, Lugaro et al. 2012, Hansen et al. 2016, Abate et al. 2018). Currently, CEMP-r stars are believed to be enriched by NSM or CC-SNe prior to their formation as they are thought to be single stars. This is, however, still under discussion, because there are only a few CEMP-r stars detected (Masseron et al. 2010). CEMP-r/s (sometimes referred to as CEMP-i) stars are enriched in both s- and r-process elements. There, the i-process may have contributed as it is proposed to be sensitive to large carbon abundances (because there, the amount of neutrons correlate with the amount of available ^{13}C). CEMP-no stars are deficient in neutron capture elements. These stars are often linked to the true second generation of stars and therefore may provide information on the first generation of stars. The analysis of CEMP-stars is challenging due to molecular bands in their spectra. An alternative approach to investigate these stars is presented in Sect. 4.4.

2.5. Chemical evolution of galaxies

The chemical composition of not only old, but also young stars, can provide information on the properties of a galaxy. The discipline of describing the evolution of chemical elements over time is called Galactic chemical evolution (GCE). The basic ingredients for such calculations are the initial conditions, the stellar birth rate function, the yields and frequency of all considered astrophysical events, and the in- and outflow of gas (see e.g., McWilliam 1997, Gibson et al. 2003, Prantzos 2008, Matteucci 2012, Nomoto et al. 2013, Matteucci 2014, Somerville and Davé 2015, for a detailed overview). The initial conditions consist of the composition of the gas and the boundary assumptions (closed or open systems with accretion and ejection of matter). The birthrate is given as the product of the star formation rate (ψ, SFR), which is the rate of stars that is born and the initial mass function (ξ, IMF), which is the mass distribution of the stars at birth. The IMF was already proposed by Salpeter (1955) to follow

$$\xi(M) \approx 0.03 \left(\frac{M}{M_\odot} \right)^{-1.35} , \tag{2.8}$$

where the negative exponent shows that our universe consists of more light than heavy stars. An indirect observable of the IMF is the ratio of a hydrostatic α-element (i.e., O, Mg, Si, S) to an explosive burning element (i.e., Ca, Ti, Fe). A higher ratio can indicate more massive progenitor stars prior to explosion (see e.g., yields of Woosley and Weaver 1995, Tsujimoto et al. 1995).

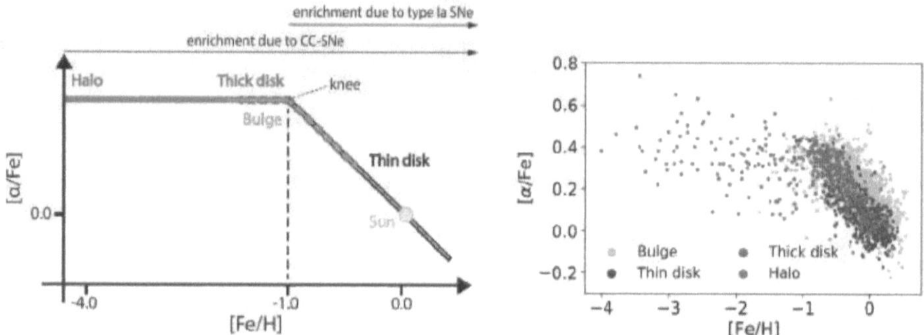

Figure 2.10.: Left panel: Schematic [α/Fe] ratio versus metallicity ([Fe/H]) of different components of the Milky Way. Adapted from Chiappini (2001). Right panel: Measured magnesium and iron abundances for different components of the Milky Way. The data was taken from Hanson et al. (1998), McWilliam et al. (1995), Zhang et al. (2009), da Silveira, C. R. et al. (2018), Johnson et al. (2014), Gonzalez et al. (2011), Bensby et al. (2013, 2014)

A schematic view of α-elemental abundances (i.e., C, O, Ne, Mg, Si, Ca, Ti) resulting from GCE calculations is shown in the left panel of Fig. 2.10. At early times (low metallicity), the Milky Way was enriched by CC-SNe that yield high [α/Fe] ratios. As mentioned above, this plateau value is sensitive to the IMF. At some point, the delayed onset of type Ia SNe causes this ratio to decrease, because these events produce large amounts of iron, but do not contribute significantly to abundances of α-elements.

In general, to trace contributions from various nuclear formation processes one can search for abundance correlations. An indication that two elements are formed by the same process can be obtained if the elemental ratio [A/B] versus [B/H] shows a flat trend and the two data sets are correlated. This indicates that elements A and B grow at the same rate (Hansen et al. 2012). Alternatively, one can use the absolute abundances of A versus B where a 1:1 ratio shows a production at the same timescales (Hansen et al. 2014a). This flat trend is also visible at early times (low metallicities) in Fig. 2.10, because α-elements and iron are both dominantly produced in CC-SNe.

The trend between α- and neutron-capture elements is still an open issue and will be discussed in chapters 5 and 6. As described above, a flat trend of these elements early on would indicate a common origin of the elements and therefore give hints on the corresponding production site of heavy elements.

2.5.1. The Milky Way

The structure of the Milky Way is quite complex and it is therefore complicated to model. It consists of substructures such as the halo, bulge, thick disk, and thin disk (Fig. 2.11) that have to be treated properly in GCE calculations. These different structures mostly show lower metallicities with increasing radius[3] (Fig. 2.10). Therefore, the oldest stars of the Milky Way are located in the halo and the sun is located towards the Galactic center in the thin disk. Moreover, the substructures differ in the kinematic properties of their stars. Halo stars can have high velocities, perpendicular to the rotational plane, whereas thick and thin disk stars tend to have smaller vertical velocities.

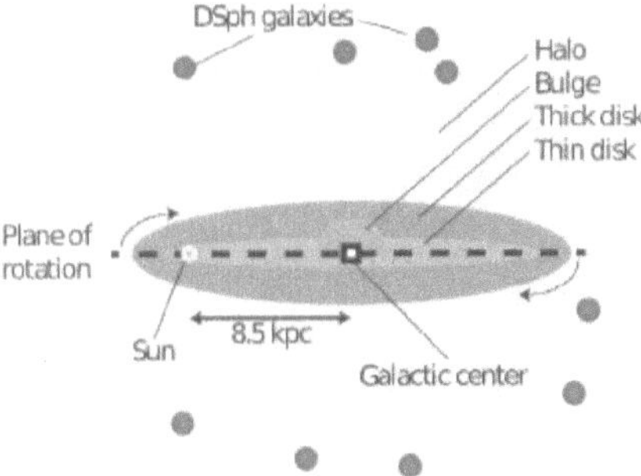

Figure 2.11.: Sketch of the Milky Way together with its satellite dwarf spheroidal galaxies. Shown are different substructures as the Galactic halo, bulge, thick disk, and thin disk. Adapted from Buser (2000).

Furthermore, the chemical signature of stars from different components may be different. In chapter 8, we investigate the origin of a peculiar star with unusual high rubidium abundance that is located in the Galactic bulge (presented in Koch et al. 2019).

---

[3]With the exception of the bulge, which hosts an old population of stars (see, e.g., Barbuy et al. 2018, for a recent review).

2.5.2. Dwarf spheroidal Galaxies

Dwarf spheroidal (dSph) galaxies are the most frequent type of galaxies in the present-day Universe. They are less massive than the Milky Way and each of the individual dSph galaxies yields information about its own unique history (a summary of their properties is given in e.g., Grebel et al. 2003, McConnachie 2012). Examples are, e.g., the large scatter of α-elements in the dSph galaxy Carina, which is thought to be a consequence of a galaxy merging event, and the kinematic properties of Segue I that indicate that this galaxy may have been a globular cluster of Sagittarius (Niederste-Ostholt et al. 2009).

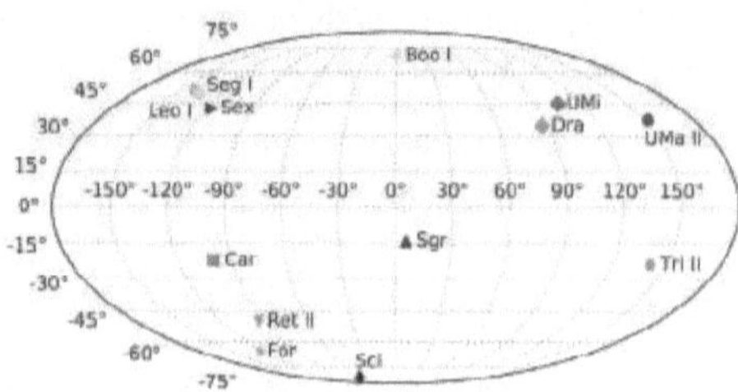

Figure 2.12.: Position of a subset of dwarf galaxies in galactic coordinates (Reichert et al. 2020).

All observed dSph galaxies so far contain a population of old stars (e.g., Grebel and Gallagher 2004), are gas-deficient, and usually found in the outskirts of massive galaxies (Fig. 2.11, see also Grebel 1999). Observing these galaxies is, however, challenging due to their faintness, but there is a number of Milky Way satellite galaxies where individual stars have been detected.

Galaxy	Abbreviation	$M_* \, [10^6 M_\odot]$	Distance[kpc]	Reference (Distance)
Sagittarius	Sgr	21	27 ± 1	Hamanowicz et al. (2016)
Fornax	For	20	146 ± 8	Karczmarek et al. (2017)
Leo I	Leo I	5.5	269 ± 12	Stetson et al. (2014)
Sculptor	Scl	2.3	86 ± 3	Pietrzyński et al. (2008)
Sextans	Sex	0.44	86 ± 6	Pietrzyński et al. (2008)
Carina	Car	0.38	106 ± 6	Karczmarek et al. (2015)
Ursa Minor	UMi	0.29	76 ± 4	Bellazzini et al. (2002)
Draco	Dra	0.29	79 ± 6	Kinemuchi et al. (2008)
Bootes I	Boo I	0.029	66 ± 2	Dall'Ora et al. (2006)
Ursa Major II	UMa II	0.0041	35 ± 2	Dall'Ora et al. (2012)
Segue I	Seg I	0.00034	23 ± 2	Belokurov et al. (2007)
Reticulum II	Ret II	-	32 ± 2	Mutlu-Pakdil et al. (2018)
Triangulum II	Tri II	-	30 ± 2	Laevens et al. (2015)

Table 2.3.: Subset of dSph galaxies together with their stellar masses taken from McConnachie (2012) and their distances.

These galaxies range from rather massive systems as e.g., Sagittarius and Fornax down to low mass galaxies such as, e.g., Reticulum II and Triangulum II. The stellar mass of a galaxy correlates with its luminosity and low-mass objects are therefore fainter. Galaxies with a magnitude of $M_V < 8$ are referred to as ultra-faint dwarf (UFD) galaxies. The stellar masses together with the typical abbreviation of some chosen dSph galaxies are shown in Tab. 2.3.

Dwarf galaxies yield a good opportunity to probe the nature of early SFH, IMF, and chemical enrichment. In chapter 5, we analyze the chemical enrichment of 380 stars of 13 dSph galaxies. This includes the values of the plateau of $[\alpha/\text{Fe}]$ (Fig. 2.10) and the position of the knee to probe the IMF and SFH, respectively.

3. Nucleosynthesis

In the previous chapter, we presented an overview of the origin and observations of elements together with the possible astrophysical hosts. This chapter will connect the occurrence of elements in stars with nuclear physics by introducing the theoretical concept of nucleosynthesis. Therefore, we discuss the nuclear cross sections and reaction rates to finally describe the synthesis of elements. We start by introducing the special case of all reactions being in equilibrium (Sect. 3.1). Afterwards, in Sect. 3.2, we derive the more general form of a set of coupled differential equations that describe the change of the composition in an astrophysical environment. The numerical methods that are necessary to solve these coupled differential equations are described in Sect. 3.3. These sections follow to large extends Cowan et al. (in preparation), Hix and Thielemann (1999), and Longland et al. (2014). Finally, we present the technical aspect of tracing the chemical evolution of an event numerically and discuss the nuclear reaction network code WinNet (Winteler 2014).

3.1. Nuclear statistical equilibrium

For high temperatures ($T \gtrsim 6\,\mathrm{GK}$ in case of explosive burning, see, e.g., Hix and Thielemann 1999) or long timescales, strong and electromagnetic forward and inverse reactions can occur frequently. Therefore, all the strong nuclear reactions can be in equilibrium, which is called nuclear statistical equilibrium (NSE). Nuclei are created and destroyed at the same rate:

$$N \cdot n + Z \cdot p \rightleftharpoons X(N, Z) + \gamma \tag{3.1}$$

with proton number Z, neutron number N, and nucleus $X(N, Z)$. Written in terms of chemical potentials, the following equation holds:

$$N \cdot \bar{\mu}_\mathrm{n} + Z \cdot \bar{\mu}_\mathrm{p} = \bar{\mu}(N, Z) := \bar{\mu}_\mathrm{N,Z}. \tag{3.2}$$

where $\bar{\mu}_\mathrm{n}$ and $\bar{\mu}_\mathrm{p}$ are the chemical potentials of neutrons and protons, respectively. To account for the change of the coulomb potential by the screening of an electron, the chemical potential $\bar{\mu}_\mathrm{n,z}$ includes a correction term that consists of

$$\bar{\mu}_\mathrm{N,Z} = \bar{\mu}_\mathrm{N,Z,no-screening} + \bar{\mu}_\mathrm{N,Z,screening} \tag{3.3}$$

with the screening correction $\bar{\mu}_\mathrm{N,Z,screening}$ (see, e.g., Chugunov et al. 2007). Screening corrections usually lead to more massive nuclei in NSE (Hix and Thielemann 1999) with stronger effects in cold environments (Lippuner and Roberts 2017). In the following, we will neglect screening corrections and assume that nuclei and nucleons obey Maxwell-Boltzmann statistics. In this case, the chemical potential is given as:

$$\bar{\mu}_\mathrm{N,Z} = k_\mathrm{B}T \ln\left(\frac{\rho N_\mathrm{A} Y_i}{G_{N,Z}} \left(\frac{2\pi\hbar^2}{m_{N,Z} k_\mathrm{B}T}\right)^{3/2}\right) + m_{N,Z}c^2. \tag{3.4}$$

Inserting Eq. (3.4) into Eq. (3.2), the NSE composition can be calculated by (see, e.g., Hix and Thielemann 1999)

$$Y(N,Z) = G_{N,Z}(\rho N_A)^{N+Z-1} \frac{(N+Z)^{3/2}}{2^{N+Z}} \left(\frac{2\pi\hbar}{m_u k_B T} \right)^{\frac{3}{2}(N+Z-1)} \exp(B_{N,Z}/k_B T) Y_n^N Y_p^Z, \qquad (3.5)$$

where $G_{N,Z}$ is the partition function that carries information of the excitation of the nucleus, $B_{N,Z}$ is the binding energy of nucleus (N, Z), $Y(N, Z)$ is the abundance, ρ is the baryon density, m_u is the atomic mass, Y_n and Y_p are the abundances of neutrons and protons, respectively. The abundance of a nucleus is defined as

$$Y_i \equiv \frac{n_i}{\rho N_A} = \frac{1}{M_i} \frac{M_i n_i}{\rho N_A} \equiv \frac{1}{M_i} X_i, \qquad (3.6)$$

where n_i is the number density of nucleus i, M_i its mass, N_A is Avogadro's constant, and the mass fraction X_i holds mass conservation

$$\sum_{i=1}^{N} X_i = 1. \qquad (3.7)$$

Furthermore, the charge is conserved

$$\sum_{i=1}^{N} Y_i Z_i = Y_e, \qquad (3.8)$$

where Y_e is the electron fraction. Equation (3.5) involves three unknowns, $Y(N, Z)$, Y_p, and Y_n. By considering Eq. (3.5), Eq. (3.7), and Eq. (3.8), the whole composition is determined by the temperature, density, and electron fraction, given that the binding energies $B_{N,Z}$ and partition functions $G_{N,Z}$ are known.

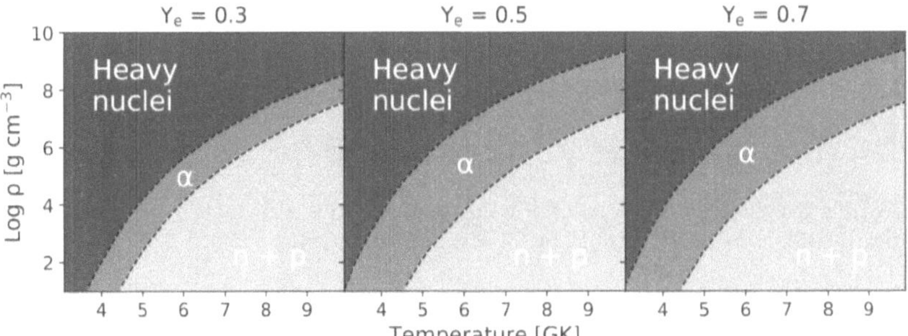

Figure 3.1.: Dominant nuclear species ($X_i > 0.5$) when assuming NSE for different temperature, density, and electron fraction. We distinguish between neutrons and protons (yellow), α-particles (green), and heavier nuclei (blue).

Higher temperatures involve photons with higher energies, which can destroy heavy nuclei by photo-disintegrations (Sect. 3.2.1) and heavy nuclei are consequently less abundant in hotter environments. This is ensured by the fact that the temperature is in the denominator of Eq. (3.5). The density behaves inversely, higher densities involve more massive nuclei, as reactions can occur more frequently. In cold, dense, or intermediate conditions, the binding energy of a nucleus becomes more relevant. The nuclei with the highest binding energy per nucleus are found in the iron group (Fig. 2.3). For intermediate conditions, due to the relatively high binding energy, α-particles get favored. The solution of Eq. (3.5) is for different conditions is illustrated in Fig. 3.1. There, we define the dominant species as the species with mass fractions of $X_i > 0.5$.

3.2. Nuclear reaction networks

When matter is cold enough, the assumption of NSE is not valid anymore and nuclear reactions become important. In the following, we will introduce the basic concept of nuclear cross sections and reaction rates to finally derive the nuclear reaction network differential equations.

3.2.1. Nuclear Cross sections and Reaction rates

A nuclear cross section contains the most basic information about any nuclear reaction. For the reaction of two particles i (target) and j (projectile)[1], the nuclear cross section is defined as the probability of the target i to react with the projectile j (e.g., Clayton 1968, Winteler 2014, Hix and Meyer 2006, Hix and Thielemann 1999)

$$\sigma = \frac{\text{number of reactions per target and second}}{\text{flux of incoming particles}} = \frac{r/n_i}{n_j v}, \tag{3.9}$$

where r is the number of reactions per cm^3 and second, n_i and n_j are the number densities of the target and projectile, respectively. The second equality holds if the relative velocity between targets and projectiles is constant and has the value v. In this case, the reaction rate can be expressed by $r = \sigma v n_i n_j$. Assuming specific velocity distributions for targets and projectiles, the reaction rate is given by

$$r_{i,j} = \int \sigma(|v_i - v_j|) \cdot |v_i - v_j| \mathrm{d}n_i \mathrm{d}n_j. \tag{3.10}$$

If the projectile is a photon, the relative velocity between projectile and target is the speed of light c and the particle distribution $\mathrm{d}n_\gamma$ is described by the Planck distribution

$$\mathrm{d}n_\gamma = \frac{8\pi}{c^3} \frac{v}{\exp(hv/k_\mathrm{B}T) - 1} \mathrm{d}v = \frac{1}{\pi^2 (c\hbar)^3} \frac{E_\gamma^2}{\exp(E_\gamma/k_\mathrm{B}T) - 1} \mathrm{d}E_\gamma. \tag{3.11}$$

For nuclei, the particle distribution $\mathrm{d}n_n$ is described by the Maxwell-Boltzmann distribution

$$\mathrm{d}n_n = n_n \left(\frac{m_n}{2\pi k_\mathrm{B}T}\right)^{3/2} \exp\left(-\frac{m_n v_n^2}{2k_\mathrm{B}T}\right) \mathrm{d}^3 v_n \equiv \phi(v_n)\, \mathrm{d}^3 v_n. \tag{3.12}$$

Therefore, for target and projectile following the Maxwell-Boltzmann distribution, we find

$$r_{i,j} = n_i n_j \int \sigma(|v_i - v_j|) \cdot |v_i - v_j| \phi(v_i)\phi(v_j)\mathrm{d}^3 v_i \mathrm{d}^3 v_j = n_i n_j \langle \sigma v \rangle_{i,j}, \tag{3.13}$$

---

[1] A nuclear reaction $i + j \to o + m$ is sometimes written as $i(j,o)m$

where $\langle\sigma v\rangle_{i,j}$ is the velocity integrated cross-section. If the projectile is a photon and the target a nucleus, the integral of Eq. (3.10) results in

$$r_{n,\gamma} = \frac{n_n}{\pi^2 c^2 \hbar^3} \int_0^\infty \frac{\sigma(E_\gamma)E_\gamma^2}{\exp(E_\gamma/k_\mathrm{B}T) - 1} \mathrm{d}E_\gamma \equiv n_n \cdot \lambda_{n,\gamma}(T). \tag{3.14}$$

Here, we introduced the decay "constant" $\lambda_{n,\gamma}(T)$. Reactions of this type are called photodisintegrations. The nuclear cross sections can be determined by experiments or approximated by theoretical models. For example, inverse reactions can be determined by assuming detailed balance (see, e.g., Iliadis 2008) and inelastic scattering of neutrons can be described by e.g., the Hauser-Feshbach model (Hauser and Feshbach 1952).

3.2.2. System of differential equations

The system of differential equations to describe the synthesis of the elements can be derived from the reactivity:

$$r_{i,j} = \frac{1}{1 + \delta_{ij}} n_i n_j \langle\sigma v\rangle_{i,j}. \tag{3.15}$$

Note that we have introduced a correction term compared to Eq. (3.13). This ensures that we do not count particles twice if the projectile and target are identical. Similarly, we can introduce the operator $\Delta_{j,k,l}$ to avoid double counting for three-body reactions

$$r_{j,k,l} = \frac{1}{1 + \delta_{jk} + \delta_{kl} + \delta_{jl} + 2\delta_{jkl}} n_j n_k n_l \langle\sigma v\rangle_{j,k,l} \equiv \frac{1}{1 + \Delta_{jkl}} n_j n_k n_l \langle\sigma v\rangle_{j,k,l}. \tag{3.16}$$

As the reaction rates are the number of reactions per time and volume, we can express the change of the number density of the species i, j, k and l that is caused by the reaction $i(j, k)l$ by

$$r_{i,j} = -\left(\frac{\partial n_i}{\partial t}\right)_{\rho=\mathrm{const.}} = -\left(\frac{\partial n_j}{\partial t}\right)_{\rho=\mathrm{const.}} = \left(\frac{\partial n_k}{\partial t}\right)_{\rho=\mathrm{const.}} = \left(\frac{\partial n_l}{\partial t}\right)_{\rho=\mathrm{const.}}. \tag{3.17}$$

Here we require that the change in number densities depends only on nuclear reactions and not on changes in the density. For the general case of many reactions that take place at the same time we obtain

$$\left(\frac{\partial n_i}{\partial t}\right)_{\rho=\mathrm{const.}} = \sum_j N_j^i r_j + \sum_{j,k} \frac{N_{j,k}^i}{1 + \delta_{jk}} r_{j,k} + \sum_{j,k,l} \frac{N_{j,k,l}^i}{1 + \Delta_{jkl}} r_{j,k,l}, \tag{3.18}$$

where N^i is the number of nuclei i that are destroyed (negative sign) or produced (positive sign) by the corresponding nuclear reaction. To get a differential equation that is insensitive to changes in the density, we express the number density n_i in terms of abundances (see Eq. (3.6))

$$\dot{Y}_i = \frac{\dot{n}_i}{\rho N_\mathrm{A}} - \frac{n_i}{\rho N_\mathrm{A}} \frac{\dot{\rho}}{\rho} \quad \text{and} \quad n_i = \rho N_\mathrm{A} Y_i \tag{3.19}$$

As a consequence, by inserting Eq. (3.19), Eq. (3.15), and Eq. (3.17) into Eq. (3.18), we obtain the ordinary differential equations

$$\dot{Y}_i = \underbrace{\sum_j N_j^i \lambda_j Y_j}_{\text{Decays and photodisintegrations}} + \underbrace{\sum_{j,k} \frac{N_{j,k}^i}{1 + \delta_{jk}} \rho N_\mathrm{A} \langle\sigma v\rangle_{j,k} Y_j Y_k}_{\text{two-body reactions}} + \underbrace{\sum_{j,k,l} \frac{N_{j,k,l}^i}{1 + \Delta_{jkl}} \rho^2 N_\mathrm{A}^2 \langle\sigma v\rangle_{j,k,l} Y_j Y_k Y_l}_{\text{three-body reactions}}, \tag{3.20}$$

where each term represents a specific type of reaction. Note that the temperature dependence of this equation is implicitly given by the reaction rates. These non-linear equations are coupled and involve an initial value problem. The initial values are usually given by either the composition of the investigated star or by assuming NSE (Sect. 3.1).

3.3. Numerical integration

To calculate the synthesis of elements, we have to numerically integrate Eq. (3.20). For a set of coupled ordinary differential equations (ODE) we can formulate the problem of this integration by the general form of:

$$\frac{\mathrm{d}y_i}{\mathrm{d}t} = \dot{y}_i = f_i(t, y_1, ..., y_N),$$

(3.21)

where N is the amount of involved nuclei and y_i the abundance of species i. There are various approaches in literature to solve these kinds of equations. However, Eq. (3.20) can involve many orders of magnitudes for different reactions and therefore being so-called "stiff". Such equations require implicit integration schemes (see, e.g., Hix and Thielemann 1999, Butcher 2008) in order to solve it. In the following, we will introduce two implicit integration schemes, the implicit Euler and Gear's method (Gear 1971).

3.3.1. Implicit Euler

The implicit Euler method (see also e.g., Hix and Thielemann 1999, Winteler 2014, Lippuner and Roberts 2017) is one of the simplest implicit integration schemes. Nevertheless, it is sufficient for most of the calculations, especially when large numbers of nuclei are involved in the calculation. By discretion of the derivative (Eq. 3.21) to

$$\frac{y_i(t+h) - y_i(t)}{h} = f_i(t+h, y_1, ..., y_N)$$

(3.22)

with time step h, we can derive an iterative formula for the solution of $y_i(t)$. This is given by:

$$y_i(t+h) = y_i(t) + h \cdot f_i(t+h, , y_1, ..., y_N).$$

(3.23)

Note that $f_i(t+h, , y_1, ..., y_N)$ is not known, because implicit methods take advantage of the evaluation of f_i at $t+h$ in order to gain numerical stability. Therefore, a root finding algorithm as the Newton-Raphson method is necessary. This method can be applied to solve equations that are formulated as:

$$\vec{F}(\vec{x}) = 0.$$

(3.24)

The Taylor series of $\vec{F}$ in first order is given by

$$F_i(\vec{x} + \delta\vec{x}) = F_i(\vec{x}) + \sum_{j=1}^{N} \frac{\partial F_i}{\partial x_j} \delta x_j + O(\delta\vec{x}^2),$$

(3.25)

where $\frac{\partial F_i}{\partial x_j}$ is one entry of the Jacobian $\mathbf{J}$ containing the partial derivatives of $\vec{F}$, defined as

$$J_{ij} = \frac{\partial F_i}{\partial x_j}.$$

(3.26)

Note that the non-linear Eq. (3.20) is linearly approximated by neglecting higher orders in Eq. (3.25). To find the root of $\vec{F}$ we apply an iteration of

$$\vec{x}^{k+1} = \vec{x}^k + \delta\vec{x}^k = \vec{x}^k - \mathbf{J}(\vec{x}^k)^{-1} \cdot \vec{F}(\vec{x}^k) \tag{3.27}$$

until convergence is reached. The convergence criteria is given by e.g., the mass conservation of the system. If

$$\left| 1 - \sum_{i=1}^{N} X_i \right| < \sigma \tag{3.28}$$

is fulfilled within a certain tolerance σ (with typical values of the order of $\sim 10^{-6}$), we assume that the algorithm has converged. This convergence criteria is sufficient for most nucleosynthesis calculations. Other criteria as $|\vec{x}_n - \vec{x}_{n+1}| < \sigma$ (for all components) are often too strict and slow down the calculation significantly (see, e.g., Lippuner and Roberts 2017). In order to integrate the ODE, one has to apply both the time integration and the root finding algorithm. By combining Eq. (3.27) and Eq. (3.23) we obtain

$$\vec{y}_{n+1}^{k+1} = \vec{y}_{n+1}^k - \left(\frac{1}{h} \cdot \mathbb{1} - \frac{\partial f(\vec{y}_{n+1}^k)}{\partial \vec{y}_{n+1}^k} \right)^{-1} \cdot \left(\frac{\vec{y}_{n+1}^k - \vec{y}_n}{h} - \vec{f}(\vec{y}_{n+1}^k) \right), \tag{3.29}$$

where the subscript denotes the implicit Euler and the superscript the Newton-Raphson iteration. Within the implicit Euler method, no error estimation is performed using higher order terms. However, we can determine the next time step h', by estimating the maximum percentage of change ϵ of the abundances, based on the current derivative. The approximate change within one time step is calculated by:

$$\left| \dot{\vec{y}}(t) \right| = \left| \frac{\vec{y}(t + h') - \vec{y}(t)}{h'} \right| \tag{3.30}$$

$$\epsilon = \max \left\{ \left| 1 - \frac{\vec{y}(t + h')}{\vec{y}(t)} \right| \right\}, \tag{3.31}$$

where the division of vectors is meant component-wise and max denotes the maximum of all components. One obtains:

$$\left| \dot{\vec{y}}(t) \right| = \left| \frac{(1 - \epsilon) \cdot \vec{y}(t) - \vec{y}(t)}{h'} \right| \Rightarrow h' = \epsilon \cdot \min \left| \frac{\vec{y}(t)}{\dot{\vec{y}}(t)} \right|. \tag{3.32}$$

with typical values of $\epsilon \approx 0.1$. In order to avoid rapid changes of the time step in our network, it is additionally limited by the previous step size

$$h' = \min \left\{ C \cdot h, \epsilon \cdot \min \left| \frac{\vec{y}(t)}{\dot{\vec{y}}(t)} \right| \right\} \tag{3.33}$$

with the constant $C > 1$ (typically $C \approx 2$). Furthermore, only species with abundances higher than a threshold abundance are taken into account in the calculation of the time step (with typical values of 10^{-10}). In order to get an adequate resolution of large temperature and density gradients, the step size is further restricted to a maximum change of temperature and density within one time step (typically $\sim 5\%$. In the case of a neutrino driven wind of a CC-SNe, tests revealed that it has to be lower $\sim 1\%$).

3.3.2. Gear's Method

In contrast to Euler's integration method, Gear's method (Gear 1971) includes terms of higher orders (see, e.g., Byrne and Hindmarsh 1975, Longland et al. 2014, Martin 2017). The method is a so-called predictor-corrector method, where in a first step a rough solution is guessed and in a second step, it is corrected until a given precision is reached. The first prediction is based on information of the past behavior of the system. Therefore, the so-called Nordsieck vector

$$\vec{z}_n = \left[\vec{y}_n, h\dot{\vec{y}}_n, \frac{h^2\ddot{\vec{y}}_n}{2!}, ..., \frac{h^q\vec{y}_n^{(q)}}{q!} \right]$$
(3.34)

is stored, where $\vec{y}_n$ are the abundances at the current time, $\dot{\vec{y}}_n, \ddot{\vec{y}}_n, ..., \vec{y}_n^{(q)}$ are the time derivatives of the abundances, h is the current step size, and q is the highest included order. The Taylor-series of $\vec{y}_n$ truncated at order q, can be calculated as the product of the Nordsieck vector with a $(q+1) \times (q+1)$ matrix defined as

$$A_{ij} = \begin{cases} 0 & \text{if } i < j \\ \binom{i}{j} = \frac{i!}{j!(i-j)!} & \text{if } i \geq j \end{cases} \qquad \text{with } i, j \in [0, 1, ..., q].$$
(3.35)

The so-called predictor step is then given as

$$\vec{z}_{n+1}^{(0)} = \mathbf{A} \cdot \vec{z}_n.$$
(3.36)

To obtain an accurate solution for $\vec{y}_{n+1}$, the predictor step is iteratively corrected due to

$$\vec{z}_{n+1} = \vec{z}_{n+1}^{(0)} + \vec{e}_{n+1}\vec{\ell},$$
(3.37)

with the correction vector $\vec{e}_n$. The $1 \times (q+1)$ vector $\vec{\ell}$ is implicitly given by

$$\sum_{j=0}^{q} \ell_j x^j = \prod_{i=1}^{q} \left(1 + \frac{(t - t_{n+1})/h}{(t_{n+1} - t_{n+1-i})/h} \right) = \prod_{i=1}^{q} \left(1 + \frac{x}{\xi_i} \right).$$
(3.38)

Here, we defined x and the vector $\vec{\xi}$ storing the information of previous step sizes

$$\xi_i = \frac{t_{n+1} - t_{n+1-i}}{h}, \qquad \text{and} \qquad x = \frac{t - t_{n+1}}{h}.$$
(3.39)

The components of $\vec{\ell} = [\ell_0(q), \ell_1(q), ..., \ell_j(q), ..., \ell_q(q)]$ are calculated as

$$\ell_0(q) = 1,$$
$$\ell_1(q) = \sum_{i=1}^{q} \left(\xi_i^{-1} \right),$$
$$\ell_j(q) = \ell_j(q-1) + \ell_{j-1}(q-1)/\xi_q,$$
$$\ell_q(q) = \left(\prod_{i=1}^{q} \xi_i \right)^{-1}.$$

To obtain the composition of the next step,

$$\left[\mathbb{1} - \frac{h}{\ell_1}J\right]\vec{\Delta}^{(m)} = -\left(\vec{y}_{n+1}^{(m)} - \vec{y}_{n+1}^{(0)}\right) + \frac{h}{\ell_1}\left(\dot{\vec{y}}_{n+1}^{(m)} - \dot{\vec{y}}_{n+1}^{(0)}\right),$$
(3.40)

$$\vec{y}_{n+1}^{(m+1)} = \vec{y}_{n+1}^{(m)} + \vec{\Delta}^{(m)}$$
(3.41)

is solved. With the number of iterations m, $\Delta^{(m)}$ as iterative correction, and **J** the Jacobian matrix, we obtain

$$J_{ij} = \frac{\partial \dot{y}_{i,n+1}^{(m)}}{\partial \dot{y}_{j,n+1}^{(m)}}.$$
(3.42)

In practice, the correction vector $\vec{e}_n$ and $\vec{y}_{n+1}$ can be derived within the same Newton-Raphson scheme. After the Newton-Raphson has converged, the correction vector

$$\vec{e}_{n+1} = \vec{y}_{n+1} - \vec{y}_{n+1}^{(0)}$$
(3.43)

can be determined. The truncation error of the abundances within one time step can be estimated by

$$\vec{E}_{n+1}(q) = \frac{1}{\ell_1}\left[1 + \prod_{i=2}^{q}\left(\frac{t_{n+1} - t_{n+1-i}}{t_{n-1} - t_{n+1-i}}\right)\right]^{-1}\vec{e}_{n+1}.$$
(3.44)

A sophisticated guess of the next time step that ensures a certain allowed tolerance ϵ (typical values of 10^{-3}, see Longland et al. 2014) can be calculated by

$$h' = hK \cdot \left(\frac{\epsilon}{\max \bar{E}_{n+1}(q)}\right)^{1/q+1},$$
(3.45)

where K is usually chosen in the interval $K \in [0.1, 0.4]$ to prevent overstepping. As for the calculation of the step size in Eq. (3.33), only abundances above a certain threshold should contribute to the calculation of the new time-step. Therefore, the truncation error is rescaled in order to prevent an overweighting of the change of very small abundances, smaller than a threshold y_{limit} (with typical values of 10^{-10}),

$$E'_{i,n+1} = \begin{cases} E_{i,n+1}/y_i & \text{if } y_i > y_{\text{limit}} \\ E_{i,n+1}/y_{\text{limit}} & \text{if } y_i \leq y_{\text{limit}} \end{cases}.$$
(3.46)

Besides the automatic control of the step size, the order q is set automatically as well. For this, we allow only order changes of $q \pm 1$. The estimated error for increasing and decreasing order are calculated by:

$$\vec{E}_{n+1}(q-1) = -\frac{\prod_{i=1}^{q-1}\xi_i}{\ell_1(q-1)}\frac{h^q\vec{y}_{n+1}^{(q)}}{q!}$$
(3.47)

$$\vec{E}_{n+1}(q+1) = \frac{-\xi_{q+1}(\vec{e}_{n+1} - Q_{n+1}\vec{e}_n)}{(q+2)\ell_1(q+1)\left[1 + \prod_{i=2}^{q}\frac{t_{n+1}-t_{n+1-i}}{t_n-t_{n+1-i}}\right]},$$
(3.48)

where Q and C are defined as

$$Q_{n+1} = \frac{C_{n+1}}{C_n}\left(\frac{h_{n+1}}{h_n}\right)^{q+1}$$
(3.49)

$$C_{n+1} = \frac{\prod_{i=1}^{q}\xi_i}{(q+1)!}\left[1 + \prod_{i=2}^{q}\frac{t_{n+1} - t_{n+1-i}}{t_n - t_{n+1-i}}\right].$$
(3.50)

To obtain the most efficient way of calculating the solution of the ODE, the step size in Eq. (3.45) is calculated for order $q - 1$, q and $q + 1$, respectively. The order providing the largest time-step is then chosen, $h' = \max(h'(q - 1), h'(q), h'(q + 1))$. For changing step sizes, the Nordsieck (see Eq. (3.34)) vector has to be rescaled:

$$\vec{z}_{n+1} = \text{diag}(1, \eta, \eta^2, ..., \eta^q) \cdot \vec{z}_{n+1},\tag{3.51}$$

where $\eta = h'/h$. The same rescaling has to be performed for decreasing orders $q - 1$. Therefore, we define a correction

$$\vec{\Delta}'_i = d_i \vec{z}_{q,n+1},\tag{3.52}$$

where, similar to Eq. (3.38), $\vec{d}$ is implicitly defined as

$$\sum_{j=0}^{q} d_j x^j = x^2 \prod_{i=1}^{q-2} (x + \xi_i)\tag{3.53}$$

and its components are given by:

$$d_0(q) = d_1(q) = 0$$
$$d_2(q) = \prod_{i=1}^{q-2} \xi_i,$$
$$d_j(q) = \xi_{q-2} d_j(q - 1) + d_{j-1}(q - 1),$$
$$d_{q-1}(q) = \sum_{i=1}^{q-2} \xi_i,$$
$$d_q(q) = 1.$$

Due to the implementation of higher orders, Gears method achieves larger step sizes compared to the implicit Euler scheme. This reduces the amount of iterations drastically without losing accuracy. However, for most calculations, more Newton-Raphson iterations are necessary, resulting in similar or even higher computational costs (Martin 2017).

Broydens method

Calculating the Jacobian is one of the most expensive steps when solving the ODE. The implicit Euler as well as Gear's method involve a root finding algorithm. There, the computation of the Jacobian of the system is involved in every iteration. Broyden (1965) developed an algorithm to correct the Jacobian iteratively instead of recalculating it in every iteration. This is done via first order corrections:

$$\mathbf{J}_n = \mathbf{J}_{n-1} + \frac{\Delta \vec{f}_n - \mathbf{J}_{n-1} \cdot \Delta \vec{Y}}{\langle \Delta \vec{Y}, \Delta \vec{Y} \rangle} \Delta \vec{Y}^T\tag{3.54}$$

with the iteration n and $\Delta \vec{Y} = \vec{Y}_n - \vec{Y}_{n-1}$ as the difference of the abundance vector, and $\langle \Delta \vec{Y}, \Delta \vec{Y} \rangle$ denotes the scalar product. Within Gear's method, the function $\vec{f}_n$ is given by the left side of Eq. (3.40). Due to rapid changes of the reaction rates and feedback from the nuclear reactions on the temperature, many more Newton-Raphson iterations are necessary to obtain convergence. Furthermore, the inverse Jacobian still has to be calculated in every step, leading to a comparable computational cost. Therefore, we neglect the usage of this method when solving the nuclear reaction network equations. We note, however, that for a system that only involves nuclear decays, Gear's method can be very efficient as the Jacobian for these problems is constant.

3.3.3. Sparse and dense matrix solvers

In principle, every nucleus is connected to every other nucleus by nuclear reactions. However, in practice most of the reactions are very unlikely and can be neglected in nucleosynthesis calculations. The most relevant reactions on a nucleus are either decays or involve the lightest nuclei, neutrons, protons, and alphas as projectile or product. In addition, there may be fission reactions that involve many potential products or the triple-alpha reaction and heavy ion reactions as e.g., $^{12}\mathrm{C}(^{12}\mathrm{C},\gamma)^{24}\mathrm{Mg}$ that are not negligible.

Most nuclei are connected to other nuclei by the reactions as shown in the left panel of Fig. 3.2. As a consequence, the Jacobian of the system (Eq. 3.26) is sparse. Figure 3.2 (right panel) shows the Jacobian of an example of 789 nuclei with a temperature of $T = 5.5\,\mathrm{GK}$ and NSE-composition (see Sect. 3.1 and cf. Hix and Thielemann 1999), where white colors indicate vanishing entries.

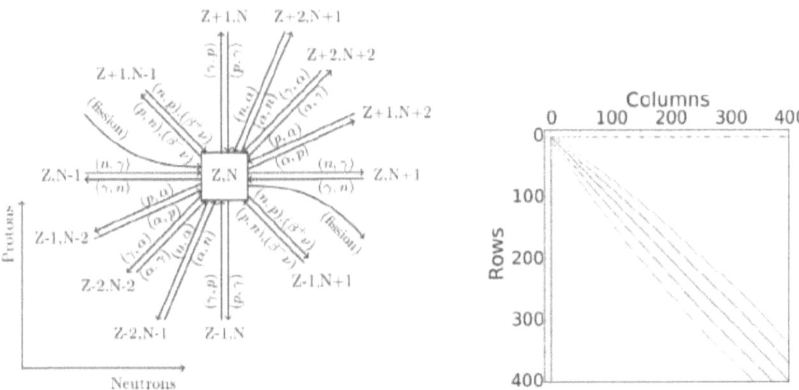

Figure 3.2.: Left panel: Sketch of the most important nuclear reactions of a nucleus. Right panel: Example of the Jacobian of a nuclear reaction network at a temperature of $T = 5.5\,\mathrm{GK}$ and NSE-composition (Reichert 2016). Red and blue colors indicate positive and negative entries, respectively. A white color indicates vanishing entries.

For sparse matrices, it can be more efficient to store only non-vanishing entries of the Jacobian. This can be done by the usage of the compressed sparse column format (cscf). To illustrate this format, we look at the matrix (see also Winteler 2014, Reichert 2016, Martin 2017)

$$X = \begin{pmatrix} 9 & 8 & 0 & 2 \\ 0 & 0 & 0 & 5 \\ 0 & 3 & 0 & 0 \\ 5 & 0 & 4 & 9 \end{pmatrix}.$$

In cscf, this matrix splits into three arrays. One array A stores the non-zero values of X:

$$A = [9, 5, 8, 3, 4, 2, 5, 9].$$

To restore the original matrix, two additional arrays have to be defined:

$$JA = [1, 4, 1, 3, 4, 1, 2, 4], \quad IA = [1, 3, 5, 6, 9],$$

where JA stores the position of the rows of A and IA the position of the columns of the non-zero entries. Clearly, a computational overhead is present when the Jacobian is converted to such format. Therefore, the sparse format will become more and more efficient as the Jacobian becomes more sparse by considering more nuclei in the reaction network. For systems that involve more than ~ 400 nuclei, the overhead is less than the advantage in performance (Fig. 3.3, see also Timmes 1999).

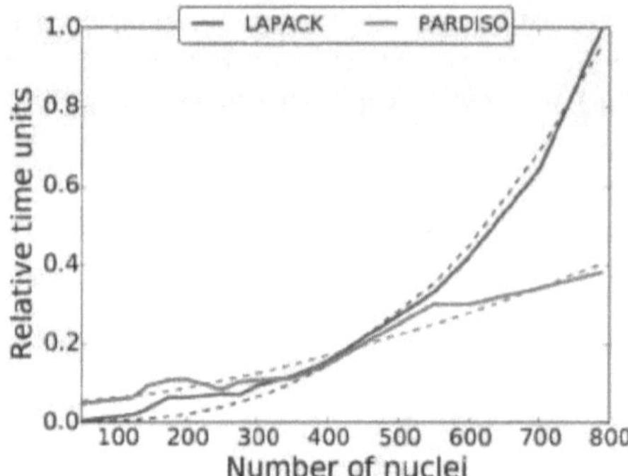

Figure 3.3.: Performance of two different nuclear reaction networks in dependence of the included amount of nuclei (Reichert 2016). Here, LAPACK (Anderson et al. 1999) denotes a solver for dense matrices and PARDISO (Intel 2016) a solver that takes advantage of sparseness. Dashed lines show a fit of the form $f(x) = a \cdot x^b + c$. Relative time units are given by the simulation time divided by the simulation time needed for a reference simulation using LAPACK and including 789 nuclei.

3.4. The reaction network code WinNet

The previous sections describe the physical and numerical basics to solve a nuclear reaction network. This section deals with the technical aspect, the concept of Lagrangian tracer particles and the implementation of the numerical methods into a reaction network code.

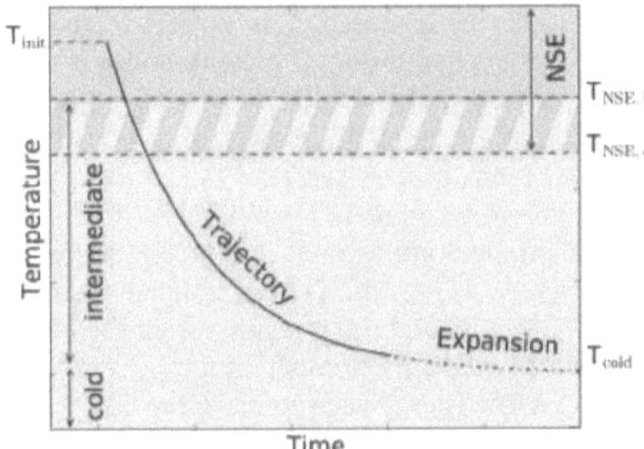

Figure 3.4.: Different temperature regimes included in WinNet together with an exemplary temperature profile of one Lagrangian tracer particle. Striped regions indicate that WinNet either assumes NSE or the intermediate regime. The transition from NSE to the intermediate regime is set by $T_{\mathrm{NSE,c}}$ and the inverse transition by $T_{\mathrm{NSE,h}}$.

In the following, we describe the nuclear reaction network WinNet (Winteler et al. 2012, Winteler 2014). It is based on the reaction network code BasNet (Freiburghaus et al. 1999, Thielemann et al. 2011), but was extended and fully rewritten in Fortran 90 (Winteler 2014). It has been used to investigate different astrophyscial conditions and nucleosynthesis processes (e.g., Winteler et al. 2012, Korobkin et al. 2012, Eichler et al. 2015, Martin et al. 2015, Bliss et al. 2018b). WinNet is a so-called single-zone code. These types of codes follow the chemical evolution of a single fluid element from a Lagrangian perspective. Hence, changes of the composition in a hydrodynamic zone are caused by nuclear burning only. This approximation is sufficient when all nuclear reaction timescales are much shorter than, e.g., convective/diffusion timescales. Limitations are e.g., Oxygen shell burning, as the convective timescales are comparable to the burning timescales (Hix and Thielemann 1999).

The time evolution of temperature, density, and neutrino properties for a fluid element can be extracted from a hydrodynamical simulation (of, e.g., CC-SNe or NSM) with Lagrangian tracer particles. The nucleosynthesis of a tracer can be post-processed by treating each particle independently (because we assume a single-zone approximation). In order to obtain the total ejected composition with sufficient resolution, a large quantity ($\sim 10^3 - 10^4$ ejected tracer for 2D simulations) of such tracer particles are included in the hydrodynamical simulation (see, e.g., Nishimura et al. 2015, Harris et al. 2017). For the integration of the total yields, the mass of each individual tracer has to be taken into account as weighting factor.

3.4.1. Code Structure

In the following, we describe the control flow of the nuclear reaction network code WinNet (see Fig. 3.5). The code starts by reading a user-defined file in the initialization step. This file contains runtime parameters such as paths to nuclear input data and other options (a full description of the different parameters is given in Appendix A). In addition, the initial time, temperature, density, and neutrino quantities are taken from the tracer.

After the initialization, the evolution mode is chosen. This mode is set to either "Network" or "NSE" and depends on the temperature. The implementation of several modes is necessary as the most efficient approach to determine the composition changes with temperature. Whereas solving the full network equations in a temperature regime where an equilibrium holds can lead to arbitrarily small time steps, solving NSE conditions in too low temperatures can lead to incorrect results. WinNet includes three temperature regimes, the regime of nuclear statistical equilibrium (NSE, see Sect. 3.1), the intermediate temperature regime, and the cold temperature regime. This is schematically shown in the right panel of Fig. 3.4. Here, two temperatures (i.e., $T_{\mathrm{NSE,h}}$ and $T_{\mathrm{NSE,c}}$) indicate the transition from the NSE to the intermediate temperature regime ($T_{\mathrm{NSE,c}}$) and the transition from the intermediate temperature to the NSE regime ($T_{\mathrm{NSE,h}}$). The need for two distinct transition temperatures is technically motivated and only necessary if feedback from nuclear energy generation to the temperature is taken into account in the calculation. In the following, we summarize the cold and intermediate temperature regimes and call them the "network regime". The cold temperature regime only differs by a replacement of reaction rates compared to the intermediate temperature regime (see Appendix A).

For both evolution modes, the temperature, density, and neutrino quantities (i.e., neutrino temperatures and luminosities) are updated using either the Lagrangian tracer particle or a user-defined extrapolation (i.e., adiabatic, exponential, free, see Appendix A). In the network regime, updating the temperature depends on the input settings and includes some special cases. If the user allows a feedback of the nuclear energy release on the temperature, a differential equation of the entropy is solved explicitly together with the nuclear reaction network equations (see, e.g., Müller 1986, Freiburghaus et al. 1999). Another special case applies to cold temperatures, as the temperature is not allowed to drop below $T = 10^{-2}$ GK. This boundary was introduced, because the included nuclear reactions (JINA database, Cyburt et al. 2010) are based on a fit that is only valid above this temperature. After updating the temperature, density, and neutrino properties, the reaction network equations are solved numerically (see Sect. 3.3.1 and 3.3.2). For the network regime, the full set of coupled differential equations (including all reactions) is solved. In the NSE regime, only weak reactions are considered in the equations and only the electron fraction is evolved. Afterwards Eq. (3.5) is used to update the composition in a user defined interval.

If no convergence is achieved, the stepsize is halved and the iteration is repeated. Otherwise, an output is generated and the time is evolved (indicated by "rotate timelevels" in Fig. 3.5). The main loop ends when an user defined termination criterion is fulfilled. Before the code terminates, final output such as the final abundances and mass fractions are written.

WinNet has been tested for reliability with simple and automated test cases that are described in Appendix B.

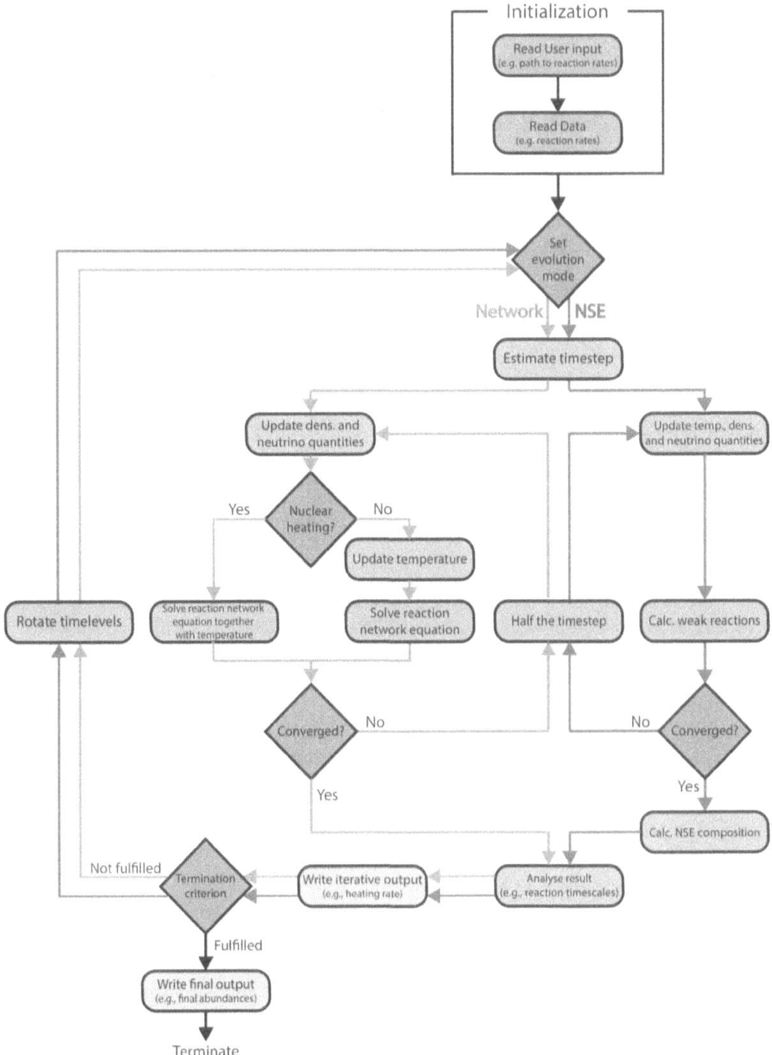

Figure 3.5.: Simplified control flow of WinNet. The code will continue until the user defined termination criterion is fulfilled. Red and blue arrows indicate the network and NSE evolution mode, respectively.

4. Chemical abundances from stellar spectra

This chapter introduces the methodology to extract abundances from stellar spectra and follows to large extends Gray (2005). We start by introducing the theoretical foundations in order to better understand the observed spectra (Sect. 4.1). Afterwards, we describe a spectrograph and recapture common methods to process a measured spectrum (Sect. 4.2). A spectrum is processed by, e.g., correcting for doppler shifts, cosmic rays, and removing sky background in order to finally be able to investigate the properties of the observed star. The derivation of stellar atmospheres as well as the stellar parameters (i.e. effective temperature, surface gravity, metallicity, and microturbulence) is described in the Sect. 4.3. This includes the description of an empirical approach to determine the metallicity of CEMP-stars that we have presented in Singh et al. (2020).

4.1. The formation of spectral lines

The light that is emitted by a star has to travel through its stellar atmosphere. On its path it is scattered and absorbed by atoms, where it excites electrons. The energy levels are different for every atom and the chemical composition therefore causes unique features in the spectrum. These features (so-called absorption lines) can be seen as dark features and contain information on the chemical composition of a star. Studying the strength and position of the absorption lines indicate the abundance of the elements that are present in the atmosphere of the star. The exact values depend on the properties of the photosphere and atomic physics quantities such as the oscillator strength and the excitation potential.

4.1.1. Local thermodynamical equilibrium and radiative transport

A photon in a stellar atmosphere can be emitted (j) or absorbed (k), their ratio is known as the source function ($S_\nu = j/k$). The source function dictates the emergent intensity and in turn the stellar flux. If all emitted photons are re-absorbed, the star is in thermodynamic equilibrium (TE), and the intensity is described by only the temperature and it radiates like a black body. This assumption may be true for the inner parts of a star where numerous collisions occur, but not in the outer photospheric parts, where more light can escape and less collisions take place.

The energy transport in the outer layers of a star is dominantly described by radiation transport. The equation of radiation transport in its general form is given by:

$$\left[\frac{1}{c}\frac{\partial}{\partial t} + (\vec{n} \cdot \nabla) \right] I_\nu = -\kappa I_\nu + j, \tag{4.1}$$

where $\vec{n}$ is the normal vector in the direction to the observer, I_ν is the specific intensity, κ is the absorption coefficient, and j the emission coefficient. For a time independent and 1D case, the transfer equation for a photon that travels in a direction s in a medium with density ρ, is given by (see, e.g., Gray 2005, for a more detailed description):

$$dI_\nu = -\kappa\rho I_\nu ds + j\rho ds. \tag{4.2}$$

By defining $d\tau = \kappa\rho ds$, we derive

$$\frac{dI_\nu}{d\tau} = -I_\nu + S_\nu. \tag{4.3}$$

A solution can be obtained by using an ansatz in the form of $I_\nu(\tau) = f\exp(b\tau)$. This leads to the integral form of the transfer equation:

$$I_\nu(\tau) = \int_0^\tau S_\nu(t_\nu)e^{-(\tau-t_\nu)}dt_\nu + I_\nu(0)e^{-\tau}, \tag{4.4}$$

where we introduced the dummy variable t_ν. The flux $\mathcal{F}_\nu$ is a related quantity to the specific intensity. For polar coordinates and without any azimuthal dependence of I_ν it can be calculated as:

$$\mathcal{F}_\nu = 2\pi \int_0^\pi I_\nu \cos\theta \sin\theta d\theta. \tag{4.5}$$

By knowing the source function $S_\nu(t_\nu)$, we can integrate the function and calculate the specific intensity as well as the flux.

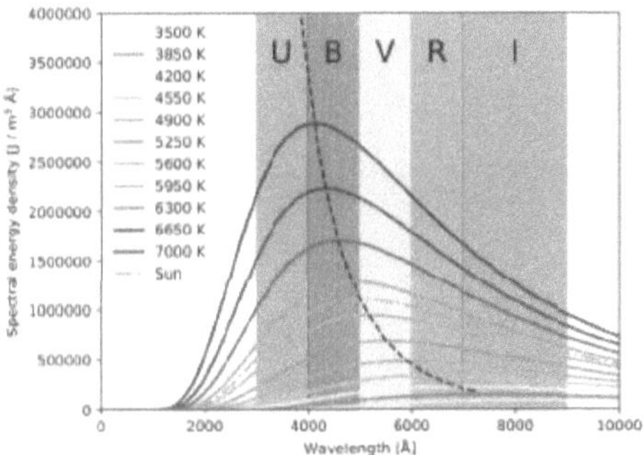

Figure 4.1.: Energy density for a black body for different temperatures. The black dashed line indicates the shift of the maximal value, described by Wien's law. The shaded regions indicate the coverage of different photometric (Johnson-Cousins) filter bands.

If you assume a local thermodynamic equilibrium (LTE) in a layer of a star, the excitation of atoms is described by the Boltzmann excitation equation, the movement by the Maxwell-Boltzmann speed distribution, and the ionization by the Saha ionization equation. In such case, the source function in one layer is given by the Planck distribution (see also Fig. 4.1):

$$S_\nu = B_\nu(\lambda, T) = \frac{2hc^2}{\lambda^5} \cdot \frac{1}{\exp\left(\frac{hc}{\lambda k_B T}\right) - 1}, \tag{4.6}$$

where h is Planck's constant, k_B the Boltzmann constant, and c the speed of light. As mentioned above, this assumption will be more precise towards inner layers of the photosphere and worse in the outer layers. This has interesting consequences as ionized atoms that are located deep inside the star will be less biased by LTE assumptions than neutral atoms. In the environment of giant stars, low densities (and surface gravities) infer lower opacities in disfavor of the LTE approximation as well (see, e.g., Lind et al. 2012, Amarsi et al. 2016, for a detailed analysis of LTE uncertainties).

4.1.2. The absorption coefficient

A photon that interacts with matter can be influenced by several factors. It can remove a previously bound electron from the corresponding atom (bound-free transition) leaving behind an ionized atom, a free electron can emit or absorb photons by the acceleration or deceleration in the presence of another charge (free-free transition), and a photon can excite a bound electron from one excitation state to another (bound-bound transition). The first two cases contribute to a continuous absorption as the wavelength of the photon is variable, whereas the latter contributes to the line absorption coefficient at a specific wavelength. The number of possible transitions of the electrons depends on the atom and its shell structure (Fig. 4.2).

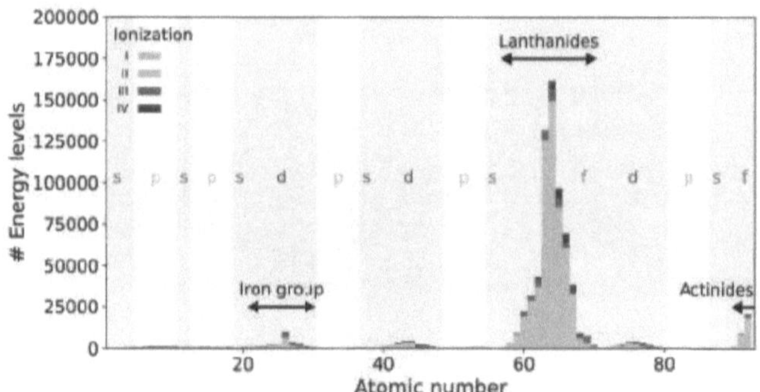

Figure 4.2.: Amount of energy levels in an atom. The data has been taken from Tanaka et al. (2019) for $Z \geq 26$ and the NIST Atomic Spectra Database (ASD, Kramida et al. 2018) for $Z < 26$.

Considering only the absorption of a photon, spectral lines should be narrow. The line will still have a finite width (upper left corner of Fig. 4.3) that is caused by Heisenberg's uncertainty principle (natural broadening). However, there are more effects that can lead to a broader absorption line. The dominant effects that shape a spectral line are natural atomic absorption, pressure broadening, and thermal Doppler broadening. Every absorption line is unique in terms of the line absorption coefficients and therefore its strength, shape, and position in the spectrum.

4.1.3. Spectral absorption lines

To fully understand and analyze a spectral absorption line, we need to know S_ν to compute the flux $\mathcal{F}_\nu$. The spectral line can be then observed as residual between a continuous flux $\mathcal{F}_c$ and $\mathcal{F}_\nu$ and expressed as

$$\frac{\mathcal{F}_c - \mathcal{F}_\nu}{\mathcal{F}_c}. \tag{4.7}$$

A spectral absorption line is therefore visible in the flux of a spectra. In reality, an immense amount of spectral lines can be present. This, however, depends on the properties and especially the metallicity of a star. While metal-poor stars have relatively few detectable absorption lines, metal-rich stars have more detectable absorption lines (see, e.g., the metal-rich giant Arcturus in Fig. 4.3).

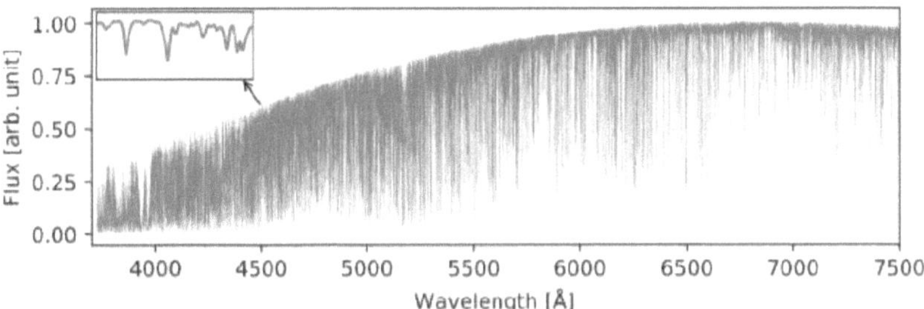

Figure 4.3.: Example spectra of the giant star Arcturus. For demonstration, we multiplied the normalized spectrum with the one of a black body. The upper left corner of the plot shows a zoomed view on the wavelength range between $4503 - 4507\,\text{Å}$.

To connect the strength of a spectral line with an abundance, we integrate Eq. (4.7) over the frequency ν:

$$EW = \int_0^\infty \frac{\mathcal{F}_c - \mathcal{F}_\nu}{\mathcal{F}_c} \mathrm{d}\nu, \tag{4.8}$$

where we defined the equivalent width (EW). The name equivalent width is derived from the fact that the area of the spectral line with normalized continuum flux can be expressed by the width of a rectangle with length one. Under LTE assumptions, the equivalent width is connected to the abundance by:

$$\log \frac{EW}{\lambda} = \log C + \log A + \log g_n f \lambda - \theta_{\mathrm{ex}} \chi - \log \kappa_\nu, \tag{4.9}$$

where C is a constant for a given ion and star, A is the abundance of element E relative to hydrogen with $A = N_E/N_H$, g_n is the statistical weight for the energy-level, $\theta_{\mathrm{ex}} = 5040/T$, and f is the oscillator strength of the atomic transition. Here, the oscillator strength is a quantity that can be measured experimentally and which describes the probability that an electron transition occurs. A higher abundance in combination with a lower oscillator strength can yield the same equivalent width. Therefore, knowing the oscillator strength precisely is important as the uncertainty directly goes (one to one) into the determined abundance.

4.2. Data processing of stellar spectra

In the last section we discussed the theoretical foundation for the description of stellar spectra. In this section, we introduce common methods and data processing when observing a real star with a telescope.

4.2.1. Spectrographs

To analyze the light of a star, a spectrograph is necessary. A typical spectrograph (Fig. 4.4) consists of an entrance slit at the focus of the telescope. Behind this slit, the photon beam diverges until it hits a collimator. Afterwards, a collimated beam propagates towards the grating. There exist different types of gratings such as, e.g., diffraction gratings, transmission gratings, reflective gratings and blazed/ruled gratings that are characterized by their design (see, e.g., chapter 3 of Gray 2005).

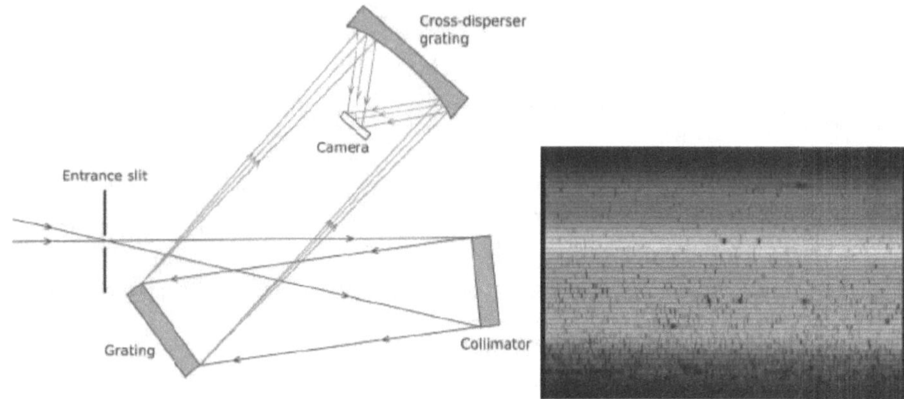

Figure 4.4.: Left: Sketch of the layout of a typical spectrograph. The light beam diverges until it reaches the collimator. Afterwards, it is reflected by two types of gratings, which send the beam to the camera that finally measures a spectrum. Right: The spectrum of the sun displayed one order on top of the other. Credit: NOAO/AURA/NSF, https://www.eso.org/public/images/sunspectrum-noao/

Due to the specific surface of gratings, the beam is separated in wavelength by constructive interference. The spectrum of the star is finally observed within a camera. A special type of spectrograph is the echelle spectrograph. There, either a cross-disperser grating or a prism is used to stack the orders on top of each other (right panel of Fig. 4.4). Due to the stacking of the orders, spectra that are observed with a echelle spectrograph usually cover a large range of wavelengths.

Another approach is to observe many stars at the same time and pipe their light via fiber optics to the entrance slit of the spectrograph. The spectrograph can be designed in a way that instead of stacking orders of one star, one order of multiple stars is stacked on top of each other (Fig. 4.5). The Fiber Large Array Multi Element Spectrograph (FLAMES, Pasquini et al. 2002) is therefore able to observe up to 132 objects simultaneously. A multi-object observation has, however, the drawback of having shorter wavelength coverage as only one row of the detector is available per observed star.

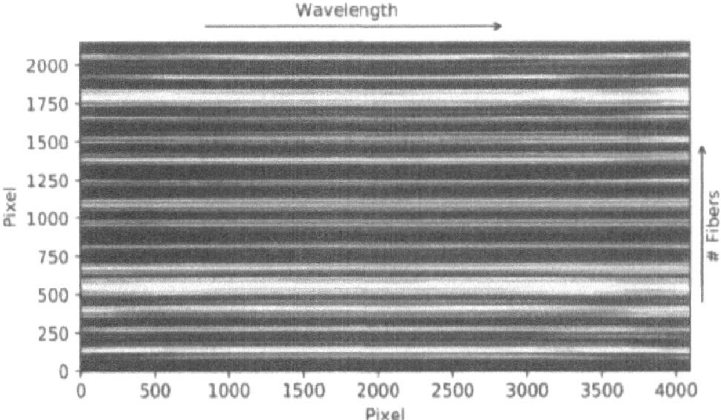

Figure 4.5.: Intensity on a CCD camera of a multi-fiber spectrograph. The wavelength is projected to the x-axis, whereas the different fibers are projected to the y-axis. The data was taken from the FLAMES/GIRAFFE spectrograph via the ESO archive[1].

---

[1] https://archive.eso.org/eso/eso_archive_main.html

4.2.2. Cosmic-rays and spectral noise

Every measurement of photons infers noise that originates in background radiation or in the components of the spectrograph. This noise is negligible for longer observation times. It is, however, not possible to observe a star infinitely long as it is restricted by the presence of clouds, the duration of the night, and the occurence of cosmic rays. Therefore, distinct observations of the same star have to be combined and co-added to reduce the spectral noise. Figure 4.6 shows the spectra of individual observations of a star in grey, and a co-added spectrum in black.

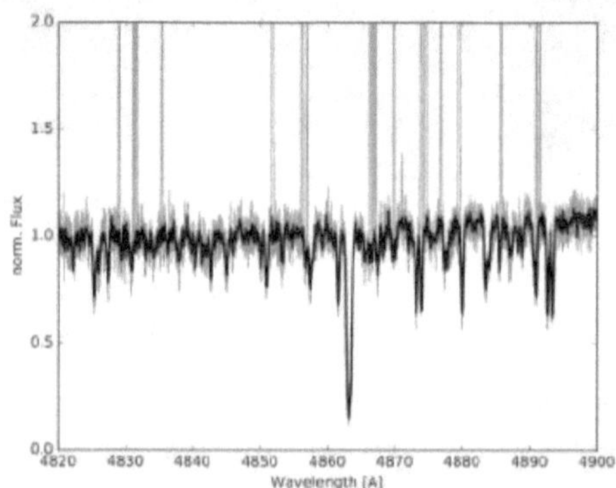

Figure 4.6.: Co-added and sigma clipped spectrum (black) together with the spectra of individual observations (grey).

In order to obtain the best signal-to-noise ratio (SNR), one can test different techniques to combine all individual observations. Depending on the method used to co-add the spectra, the final SNR will differ. For tests with synthetic spectra, the best SNR was obtained by co-adding the spectra using the square weighted average:

$$\mathcal{F} = \frac{\sum \left(\mathrm{SNR}_i^2 \cdot \mathcal{F}_i\right)}{\sum \mathrm{SNR}_i^2}, \tag{4.10}$$

with $\mathcal{F}$ as the resulting co-added flux, and $\mathcal{F}_i$ the flux of the individual spectra i. It is worth noting that the SNR of a measurement strongly depends on the brightness of the target. For distant objects, this infers observational biases as usually only bright giants are observed, because they can reach the desired SNR faster than fainter dwarf stars.

Another uncertainty in the observation of a star is caused by cosmic rays. This is visible in the spectrum shown in Fig. 4.6 as strong and narrow peaks. Therefore, a sigma clipping must be performed so that the final spectrum has no more spikes.

4.2.3. Sky contamination

Most telescopes are ground-based. This means that light from an observed star has to travel through
Earth's atmosphere and the interaction of the photons will be visible as absorption or emission lines in
the measured spectrum. Furthermore, there will be a continuous background flux that does not originate
from the star. Therefore, every observation of a star has to be done simultaneously with a measurement
of the sky to be able to finally account for this effect. For an observation of stars (blue circles in Fig. 4.7)
with FLAMES (Pasquini et al. 2002), there are always fibers pointing towards places on the sky, where
no star is visible (see blue crosses in Fig. 4.7).

Figure 4.7.: Example of a multi-fiber spectrograph observation. Stars are shown as black dots. Blue
circles indicate science targets, whereas sky-fibers are shown with blue crosses. Adopted
from European Southern Observatory (2009).

Subtracting the sky signal from the spectra of the star can also introduce noise. Therefore, a previously
performed median over all sky-fibers can create a so-called master-sky spectrum with a lower SNR.
However, for the subtraction of the sky-background various methods exist as, e.g., the scaling of the sky
by a factor that is determined from spectral features (see, e.g., Hill et al. 2019) or a straight average of
all sky-fibers (e.g., chapter 5).

Besides a continues absorption/emission there may be spectral emission and absorption lines from, e.g.,
oxygen of Earth's atmosphere. An analysis of a spectral line at the same position as an atmosphere
feature (tellurics) can be very uncertain. It is possible to correct for tellurics, however, there will remain
a reduced number of photon counts, which will increase the noise at this part of the spectrum.

The atmosphere of the Earth also absorbs most of the ultra-violet light. Some elements (e.g., Sn or second
r-process peak elements), however, have their strongest absorption lines in this region of the spectrum.
As a consequence, these elements can not be observed with ground-based telescopes (see the gap at
the second peak in Fig. 2.9). This is one of the reasons that telescopes that are located outside of the
atmosphere as the Hubble Space Telescope (HST) exist.

4.2.4. Velocity determinations

The absorption lines are always located at the same wavelength if measured in the lab frame. However, if the star moves with respect to the observer, the spectrum must be corrected for the Doppler effect. The resulting shift in wavelength $\Delta\lambda$ is given by:

$$\Delta\lambda = \frac{\lambda_0 c}{v_r}. \tag{4.11}$$

This effect makes it possible to calculate the radial velocity v_r of a star by comparison to a reference spectrum at rest frame. Note that this implies that telluric absorption lines will finally end up at a different wavelength. In addition, proper motions can be observed by measuring the angular change of the star on the sky.

The velocity of a star is dominantly influenced by the velocity of the system to which it belongs. Therefore, by comparing radial velocities as well as proper motions of a star with the ones of globular clusters, dwarf galaxies, or other systems one has a powerful tool to assign membership to a particular system. This has the advantage that the determination of the velocity is more accurate than other methods, such as measuring the distance of a star.

4.3. Stellar parameter and atmosphere models

In order to calculate the radiative transport of a photon, one has to model the photosphere. In practice, most models are calculated in 1D. They are calculated on a grid that consists of quantities such as temperature, density, pressure, opacity, and abundance of electrons tabulated at different optical depths. The generation of an atmosphere model depends on the stellar parameters, i.e., effective temperature, surface gravity, metallicity, and microturbulence. The calculation of atmosphere models can be quite challenging as they often do not converge. Therefore, instead of calculating a new model, they are often interpolated from pre-computed models on a grid of the stellar parameters. In the following, we will introduce methods to derive the stellar parameters.

4.3.1. Effective temperature

There are multiple methods and approaches to derive the effective temperature of a star. Here, we will summarize three common methods.

Reminding ourselves that a star can be approximately described as a black body, there exists a relation between the effective temperature and the color of a star. This method is known as the infrared flux method (IRFM). The color can be defined by the difference of magnitudes in different filter bands (see Fig. 4.1, for a simplified view of the wavelength coverage of the filter system). A detailed overview of the individual filter systems, including their conversion equations is given in Bessell (1979). However, finding a relation for the effective temperature of stars is more complex than just assuming a black body, because the absorption lines of the atoms can also influence the magnitude. Therefore, empirical approaches are performed to relate the color and metallicity to the effective temperature of the star (e.g., Blackwell and Shallis 1977, Alonso et al. 1996, 1999, Ramírez and Meléndez 2005, Casagrande et al. 2006, González Hernández and Bonifacio 2009, Casagrande et al. 2010). The correlation between B-V colors and effective temperatures is shown in the left panel of Fig. 4.8 (with data taken from Casagrande et al. 2010). Note that the individual colors have to be corrected for the dust between the star and the observer (taking, e.g., values from the infrared science archive, IRSA). Shorter wavelengths have a higher cross-section (caused by absorption and scattering) compared to larger wavelengths on the dust particles and the colors are therefore shifted to redder values. Correcting for this effect is referred to as dereddening.

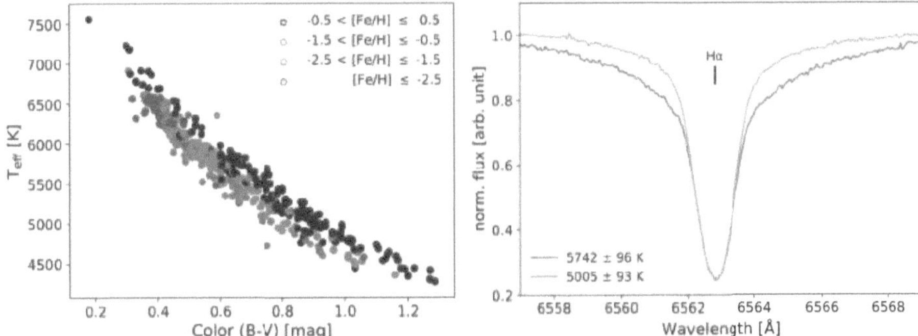

Figure 4.8.: Left panel: Correlation between the effective temperature and the (B-V) color. The data for this plot was taken from Casagrande et al. (2010). Right panel: Temperature dependent absorption line of the stars CS 31065-0008 (blue) and HE 1430+0053 (orange). Here, the effective temperatures where calculated by the code ATHOS (Hanke et al. 2018).

Another approach to determine the effective temperature is the investigation of the shape of a temperature dependent absorption line. Ideal candidates are the strong absorption lines of the Balmer series Hα (located at 6562.8 Å), Hβ (located at 4861.3 Å), and Hγ (located at 4340.5 Å). Determining the shape of these lines is equivalent to determining the effective temperature of the star. Lower temperatures indicate a slower movement of the atoms, which results in less broadening and in turn damping. Besides other damping effects, the Doppler broadening of the line is weaker in comparison to larger temperatures. Therefore, higher temperatures have a broader Hα, Hβ, and Hγ absorption line (see right panel of Fig. 4.8). This effect can be used to correlate the shape of the line with the effective temperature (see e.g., Barklem et al. 2002, Lind et al. 2008, Hanke et al. 2018). Note that the shape of the balmer lines are highly sensitive to 3D effects (Amarsi and Barklem 2019) and can therefore not be modeled properly in 1D.

The effective temperature can also be determined in an iterative process. For this, it is necessary to measure the EW of several Fe I lines. In combination with the stellar atmosphere model, every line will lead to an individual abundance. These abundances have to be consistent among each other. Colder effective temperatures will lead to less excitements of the absorption lines and in fact lower abundances for a fixed EW. On the other hand, higher temperatures will excite more atoms and therefore lead to higher abundances (Fig. 4.9). The strength of this effect can vary and it depends on the excitation potential of the absorption line (e.g., Fig. 16.2 of Gray 2005). Absorption lines with low excitation potentials are more affected than the ones with higher excitation potential. This leads to a negative slope when plotting abundances versus excitation potentials in the case of higher effective temperatures and to a positive slope in the case of low effective temperatures.

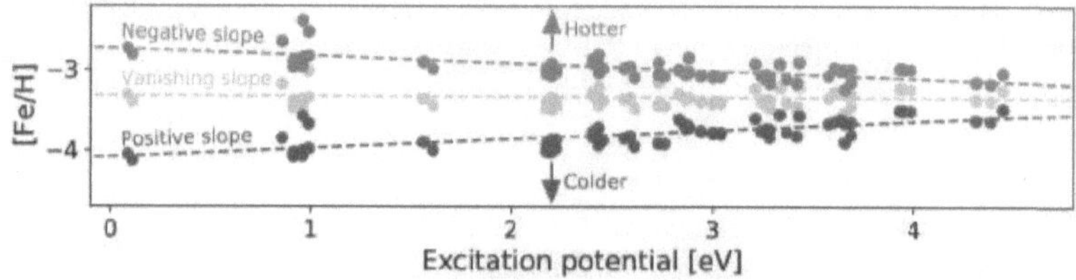

Figure 4.9.: Metallicity in dependence of the excitation potential for the star HE 1430+0053. Every dot in a certain color indicates a different Fe I absorption line. Shown are the derived (LTE) metallicities for different Fe I absorption lines assuming an atmosphere model with $T_{\text{eff}} = 5200\,\text{K}$ (red), $4800\,\text{K}$ (orange), and $4400\,\text{K}$ (blue).

A vanishing slope states that the model temperature is consistent for all measured absorption lines. Note that the determination of the temperature has to be done simultaneously with the determination of all other stellar parameters and there can be slight effects on the slope from the surface gravity, metallicity, and microturbulence.

4.3.2. Surface gravity

Similar to the effective temperature, there are multiple methods to determine the surface gravity. Consider the following relations (see also, e.g., Nissen et al. 1997)

$$g \propto \frac{M}{R^2} \text{ and } L_\star \propto R^2 T_{\text{eff}}^4, \tag{4.12}$$

where M is the stellar mass, R the stellar radius, and L the luminosity of the star. The second relation is known as Stefan-Boltzmann equation applied on a black body. The luminosity in terms of a bolometric magnitude M_{bol} is given by:

$$M_{\text{bol}} = -2.5 \log \frac{L_\star}{L_0} = M_V + BC + 5 \log \pi + 5, \tag{4.13}$$

where L_0 is a zero-point luminosity that is arbitrary defined as $L_0 = 3.0128 \cdot 10^{28}\,\text{W}$, BC is the bolometric correction, M_V is the absolute visual magnitude of the star, and π is the parallax in arcseconds. By combining Eq. (4.12) and Eq. (4.13) and in addition normalizing it to the sun, we obtain

$$\log \frac{g}{g_\odot} = \log \frac{M}{M_\odot} + 4 \log \frac{T_{\text{eff}}}{T_{\text{eff},\odot}} + 0.4 \cdot (M_{\text{bol}} - M_{\text{bol},\odot}) \tag{4.14}$$

$$= \log \frac{M}{M_\odot} + 4 \log \frac{T_{\text{eff}}}{T_{\text{eff},\odot}} + 0.4 M_V + 0.4 BC + 2 \log \pi + 0.12. \tag{4.15}$$

As a consequence, the knowledge of the distance, the effective temperature, and the stellar mass can be used to calculate the surface gravity. The stellar mass is difficult to measure and therefore often assumed as $0.8M_\odot$. This assumption is not problematic as Eq. (4.15) dominantly depends on the parallax of the star. A variation in the mass has only minor effects on the derived surface gravity.

Another way to determine surface gravities can be obtained by the position in the HR-diagram (see Fig. 4.10), and by knowing the age and metallicity of the star (e.g., Demarque et al. 2004). The usage of isochrones to determine the surface gravity can be beneficial when the stellar spectrum is noisy and the distance is unknown.

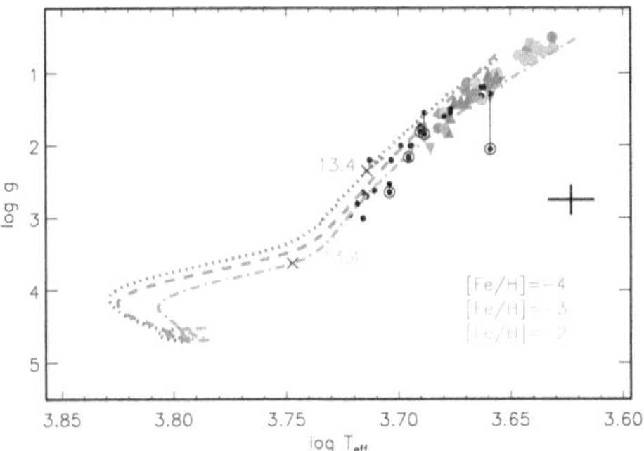

Figure 4.10.: Stars together with different theoretical evolutionary tracks in the HR-diagram (Mashonkina et al. 2017a).

The absorption features in the spectrum of a star also yield information of the surface gravity. Constrained by the Saha ionization equation, both the neutral Fe I and the ionized Fe II should yield the same overall Fe abundance because we assume that all elements are uniformly mixed in the stellar atmosphere. A mismatch can be cured by changing the surface gravity of the model. An increasing surface gravity will result in an higher Fe II abundance, whereas a lower surface gravity will give a lower Fe II abundances (Fig. 4.11).

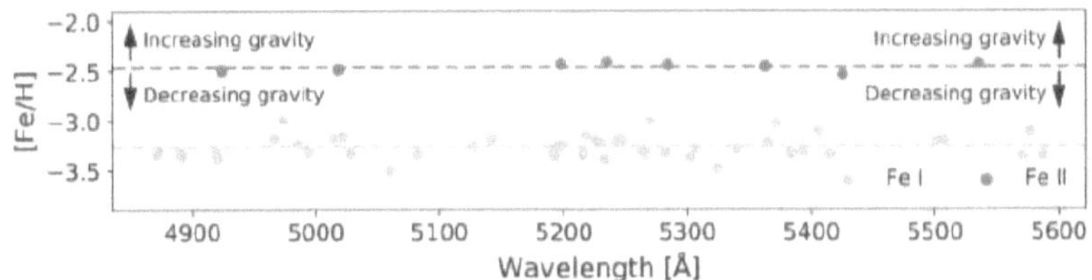

Figure 4.11.: Metallicity for Fe I lines (orange) and Fe II lines (green). A lower surface gravity results in a lower Fe II abundance. This example has been taken from the star HE 1430+0053.

The neutral Fe I only weakly depends on the surface gravity. However, we have to keep in mind that there are larger biases due to the LTE assumption on Fe I abundances, which are formed in the outer layers of the star, where a black body becomes a poorer approximation. Depending on the metallicity, this leads on average to too low Fe I abundances (see e.g., Lind et al. 2012). As a consequence, forcing excitation equilibrium of Fe I and the less biased Fe II results in too low surface gravities.

In addition, Fe II lines tend to be weaker than Fe I lines. Therefore, less of these lines can be detected and the number statistics can be rather low, especially for metal-poor stars.

4.3.3. Microturbulence

The microturbulence ν_T is defined as small-scale motions that are small compared to the mean free path of a photon. The microtubulence cause (similarly to the thermal case) a gaussian line broadening. It influences strong lines different compared to weak lines. Therefore, it can be determined by forcing a vanishing slope of the Fe I or Fe II abundances versus the EW (Fig. 4.12). The parameter is included to compensate for the lacking 3D convective motions that cannot be modeled in 1D (e.g., chapter 17 of Gray 2005).

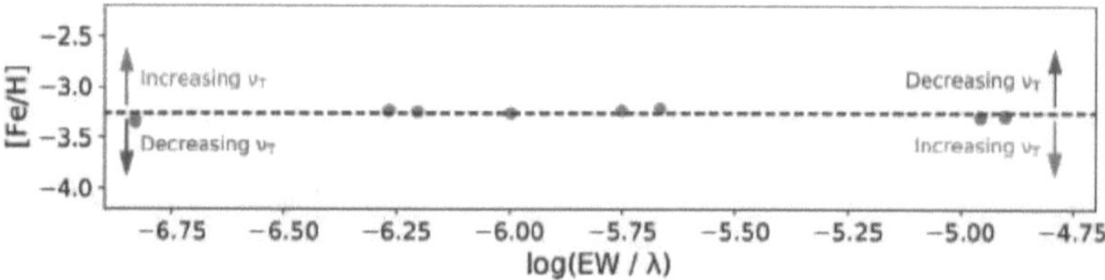

Figure 4.12.: Logarithm of the equivalent width divided by the wavelength versus metallicity for eight Fe II absorption lines for the example star HE 1430+0053.

As it is strongly correlated with the stellar parameters of a star, multiple empirical formulae have been developed to compute ν_T. Kirby et al. (2009) (and similarly Marino et al. 2008) presented the empirical formula:

$$\nu_T = ((2.13 \pm 0.05) - (0.23 \pm 0.03) \cdot \log g)\,\mathrm{km\,s^{-1}}. \tag{4.16}$$

Another relation that involves also temperature and metallicity was presented by Mashonkina et al. (2017a):

$$\nu_T = (0.14 - 0.08 \cdot [\mathrm{Fe/H}] + 4.90 \cdot (T_\mathrm{eff}/10^4) - 0.47 \cdot \log g)\,\mathrm{km\,s^{-1}}. \tag{4.17}$$

These relations can be used whenever few Fe lines are available in the stellar spectra.

4.3.4. Metallicity

The metallicity of the star is approximately connected to the time of its formation (see Sect. 2.5) and therefore an indicator of the general composition. It can be determined by measuring Fe I or Fe II lines and knowing all other stellar parameters. Figure 4.13 shows the absorption lines of three Fe I lines of the star CS 31065-0008 together with three synthetic spectra assuming a metallicity of -3.10 dex, -3.40 dex, and -3.86 dex. There, a clear difference in the strength of the absorption lines is visible. Metal-poor stars have weaker lines than metal-rich ones.

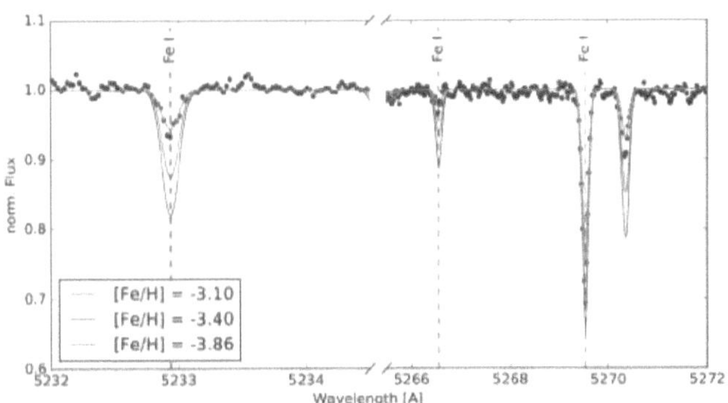

Figure 4.13.: Three Fe absorption lines of CS 31065-0008 together with three synthetic spectra using model atmospheres with different metallicities (see legend).

An estimate of the metallicity can be also extracted by over-plotting stars with similar effective temperature, surface gravity, and microturbulence and known metallicity. Instead of taking spectra of stars, a grid of previously calculated synthetic spectra can be used to determine the metallicity by minimizing the residual between the synthetic and observed spectra (Kirby et al. 2010, Boeche and Grebel 2016).

4.4. Constraining metallicities of CEMP-stars by empirical formulas

Many metal-poor stars are C-rich (Sect. 2.4.1) and it is crucial to have a way to determine the metallicities of these stars in an accurate way. This can be a particular challenge, because large carbon abundances cause molecular bands in the spectra. These bands can be easily mistreated and lead to a wrong continuum placement and, as a consequence, to wrong metallicities exceeding an uncertainty of 0.25 dex. Especially when observing with low resolution, many lines can blend and smear out, making it even more complicated to extract a metallicity.

Within Singh et al. (2020), we investigated an empirical method to determine the metallicity of CEMP-stars that are observed in low-resolution. In the following, we will recapture the findings of this work.

We correlated the equivalent widths (EW) of strong spectral features, which can be identified also in metal-poor stars, with the metallicity of the star. Equivalent width do not rely on prior knowledge of the stellar parameters and are therefore not biased by 1D or LTE assumptions (see Sect. 4.1.1), however, the underlying metallicities of the correlation do rely on 1D and LTE assumptions. Our sample consists of 82 (CEMP and C-normal) stars. Parts of our sample are well known and observed at high resolution, whereas others are poorly-known stars observed at low and medium resolution.

A similar relation is presented in Wallerstein et al. (2012). There, they used the EW of a Ca II line at $\lambda = 8498$ Å to correlate it with metallicities of the variable RR Lyrae stars. Their linear relation is given by:

$$[\mathrm{Fe/H}]_{\mathrm{RRLyr}} = -3.846(\pm 0.155) + 0.004(\pm 0.0002) \cdot \mathrm{EW}_{\mathrm{Ca}}. \tag{4.18}$$

Our aim was to find a similar relation that is suitable for CEMP-stars, because using Eq. 4.18 leads for our CEMP-stars to large uncertainties up to ± 0.4 dex. We explored several spectral features, which have a residual standard deviation that is better than 0.25 dex for stars with metallicities $[\mathrm{Fe/H}] < -2$. These spectral features have been preselected based on a cut in the oscillator strength ($\log gf \gtrsim -2$) and excitation potential (EP $\lesssim 3$ eV) to ensure that they are strong and detectable in metal-poor giant as well as dwarf stars. We determined the EWs of elements for Na I (27 stars), Mg I (28 stars), Ca I (9 stars), Ca II (27 stars), Sc II (22 stars), Cr I (44 stars), Mn I (12 stars), and Ni I (39 stars). By using spectral synthesis, we were also able to determine the elemental abundances of C (CH) (30 stars), Mg I (30 stars), Sr II (30 stars), Ba II (29 stars), and Eu II (8 stars). These abundances are used to classify the CEMP stars (see Table 2.2).

The Mg I triplet is strong and detectable in the spectra of most stars (at our spectral resolutions and metallicities). However, large molecular bands at $\lambda \sim 5167$ Å makes it unusable for our purpose. A linear relation of the Mg lines at $\lambda = 5173$ Å and 5184 Å shows a residual standard deviation of ± 0.41 dex and ± 0.35 dex, respectively.

The Cr I lines at 5844 Å, and 7400 Å, Mg I at 6173 Å, Mn I at 4762 Å, and Sc II lines at 5239 Å and at 6605 Å, show a nearly flat relation with a large scatter. This indicates that the EWs are degenerate with metallicity. Furthermore, their residual standard deviations exceed ± 0.40 dex, which does not fulfill the targeted accuracy of 0.25 dex. The Na I lines at 5890 Å and 5896 Å, Mg I at 8806 Å and Ca I at 6162 Å show a clear linear relation, however with a large scatter that amounts to residual standard deviations exceeding ± 0.44 dex.

Suitable relations are derived from the Cr I triplet located at ~ 5204 Å. An exception is the absorption line at 5206 Å as this line is heavily blended with Y II (see Fig. 4.14). The absorption feature of Y II varies for CEMP-s/r compared to CEMP-normal stars. Our aim was to find an absorption feature that behaves similar in all CEMP stars and we therefore exclude this line from our analysis. We find that by adding the EWs of Cr I at 5204 Å and 5208 Å we obtain the following linear relation:

$$[\mathrm{Fe/H}]_{\mathrm{Giants}} = -3.022(\pm 0.066) + 0.002(\pm 0.0002) \cdot \mathrm{EW}_{\mathrm{Cr(5204+5208)}}, \tag{4.19}$$

where EW is given in mÅ. For our sample, this relation yields a residual standard deviation of 0.25 dex including CEMP as well as C-normal giant stars. Similarly we find

$$[\mathrm{Fe/H}]_{\mathrm{Dwarfs}} = -2.565(\pm 0.071) + 0.005(\pm 0.0003) \cdot \mathrm{EW}_{\mathrm{Cr(5204+5208)}} \tag{4.20}$$

for dwarf stars with a residual standard deviation of 0.25 dex.

Another strong spectral feature is a Ni I absorption line at 5477 Å. Even when blended with Fe (Fig. 4.14), it does not degrade the use of this Ni line. We derive

$$[\mathrm{Fe/H}]_{\mathrm{Giants}} = -2.965(\pm 0.058) + 0.005(\pm 0.0003) \cdot \mathrm{EW}_{\mathrm{Ni(5477)}}, \tag{4.21}$$

which involves a residual standard deviation of $0.22\,\mathrm{dex}$. This makes the feature a better tracer of the metallicity than conventional methods and the most accurate relation within our investigation. For dwarfs, the linear relation is given by

$$[\mathrm{Fe/H}]_{\mathrm{Dwarfs}} = -2.603(\pm 0.052) + 0.015(\pm 0.001) \cdot \mathrm{EW}_{\mathrm{Ni}}. \tag{4.22}$$

Here, the residual standard deviation is $0.19\,\mathrm{dex}$ and even more accurate than for giants.

From a GCE point of view, Ni is an understandable tracer of the metallicity as [Ni/Fe] shows a flat trend and a low scatter (0.03 for very metal-poor stars, Reggiani et al. 2017) with metallicity (e.g., Kobayashi et al. 2019). Cr, on the other hand, shows a decreasing of [Cr/Fe] below [Fe/H] $= -2.5$ (Bonifacio et al. 2009). This trend may be however an effect of LTE assumptions (see e.g., Bergemann and Gehren 2008, Mashonkina et al. 2017a), which is supported by the less LTE biased [Cr II/Fe] abundances (Bonifacio et al. 2009). Therefore, the increasing trend of [Cr I/Fe] may not propagate into the EW of the absorption feature, but only be an artefact of the LTE model assumption.

Hence, selecting the two Fe-peak elements (i.e., Cr and Ni) to trace the metallicity, seems logical from a nucleosynthetic point of view. This technique and development of an empirical metallicity tracer will be useful in future surveys where metallicities of millions of stars need to be accurately and precisely derived.

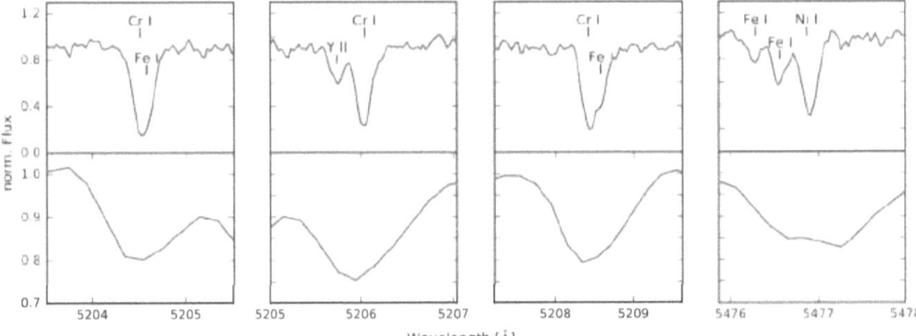

Figure 4.14.: Cr and Ni lines in the high-resolution spectrum of the Sculptor star 2MASS J01001255-3343011 ([Fe/H] $= -1.62$, top) and the low-resolution spectrum of the sample star HE 0253-6024 ([Fe/H] $= -2$, bottom). This plot was taken from Singh et al. (2020).

Therefore, massive tests in all kind of stars have to be performed in order to estimate a realistic error. In the next chapter, we will perform tests with a large sample of giant stars. Tests with dwarf stars have also been performed and outlined in Singh et al. (2020).

5. Neutron-capture elements in dwarf galaxies

In the following, we analyze dSph member stars from the public European Southern Observatory (ESO[1]) and Keck[2] reduced archives. First, we present the stellar parameters of these stars, using as few methods as possible to derive them. This is necessary to ensure homogenization of the abundances over our entire sample. Si=smallnce our sample mainly consists of cool giants that have similar stellar parameters compared to CEMP-stars, we use the derived stellar parameters to test the empirical relations, Eq. (4.19), and Eq. (4.21), which were outlined in Sect. 4.4 and presented in Singh et al. (2020). We then derive abundances and investigate possible trends between heavy neutron-capture elements and α-elements.

This chapter was to large extends already presented in Reichert et al. (2020) and was a collaborative work together with C. J. Hansen, M. Hanke, Á. Skúladóttir, A. Arcones, and E. K. Grebel. Section 5.3 was presented in Singh et al. (2020) in collaboration with D. Singh, C. J. Hansen, J. S. Byrgesen, and H. M. Reggiani.

5.1. Sample selection

We initially selected 5497 stars for the analysis, which were previously identified as members of dSph and UFD galaxies. This also includes stars of dSph galaxies from the SAGA database. The sample contains high-resolution data ($R \gtrsim 15000$) that were subject of previous studies, but also includes unpublished data. The reduced ESO archive and the Keck Observatory archive were explored and we selected spectra covering the ultraviolet and visible to include elemental transitions that are key for the present study. We did not include stars that only contain the infrared part of the spectrum used for Ca triplet (CaT) surveys (e.g., Pont et al. 2004, Battaglia et al. 2006, Koch et al. 2006, Hendricks et al. 2014b). Stars for which we could not determine stellar parameters (i.e., effective temperature, surface gravity, metallicity, and microturbulence; Sect. 4.3) and stars with insufficient data (e.g., S/N < 10) or missing spectra were removed from the analysis. For example, we did not include stars from McWilliam and Smecker-Hane (2005) that are found in the Keck archive, as we could not accurately (within a 3 arcsec circle) resolve their coordinates in the SIMBAD[3], NED[4] or the *Gaia* archive (Gaia Collaboration 2018b). In addition, we determined radial velocities (Sect. 4.2.4) of each star to confirm membership in the individual galaxies. Stars analyzed in Tafelmeyer et al. (2010) were added manually to include coordinate corrections from Tafelmeyer et al. (2011).

---

[3]Set of Identifications, Measurements and Bibliography for Astronomical Data, Wenger et al. 2000

[4]The NASA/IPAC Extragalactic Database (NED) is operated by the Jet Propulsion Laboratory, California Institute of Technology, under contract with the National Aeronautics and Space Administration.

Table 5.1.: Number of stars per galaxy in our sample. In addition, the median SNR at 5528 Å per resolution element and dSph galaxy is given (Reichert et al. 2020).

Galaxy	Abbreviation	SNR	# Stars
Sagittarius	Sgr	141	32
Fornax	For	79	123
Leo I	Leo I	52	2
Sculptor	Scl	93	96
Sextans	Sex	86	46
Carina	Car	47	47
Ursa Minor	UMi	88	13
Draco	Dra	92	7
Bootes I	Boo I	38	3
Ursa Major II	UMa II	98	3
Segue I	Seg I	300	1
Reticulum II	Ret II	47	6
Triangulum II	Tri II	173	1
Total			380

This resulted in a total sample size of 380 stars in 13 dwarf galaxies (see Table 5.1 and Fig. 2.12), with 295 observations from FLAMES/GIRAFFE (Fibre Large Array Multi Element Spectrograph, Pasquini et al. 2002, with resolving powers ranging from 17000 to 28800), 56 from UVES (Ultraviolet and Visual Echelle Spectrograph, Dekker et al. 2000, with resolving powers of 31950 and 42310), 2 from X-shooter (Vernet et al. 2011, with resolving powers of 6600 and 11000), and 27 from HIRES (High Resolution Echelle Spectrometer, Vogt et al. 1994, with resolving powers of 35800, 47700, and 71600). The final sample size is dictated by the accessibility of the spectra obtained with HIRES, UVES, FLAMES/GIRAFFE or X-shooter in the reduced archive. Unfortunately, observatories of other high-resolution instruments do not provide archives with reduced spectra.

5.2. Homogenized stellar parameter

Homogenized stellar parameters are a key to understand abundance trends and variations among different dSph galaxies. Differences in the analysis of stellar spectra can directly influence the derived abundances (see, e.g., Fig. 5.1). Therefore, we tried to determine the stellar parameters as homogeneous as possible. In the following, we will outline the methods that we have used for our sample.

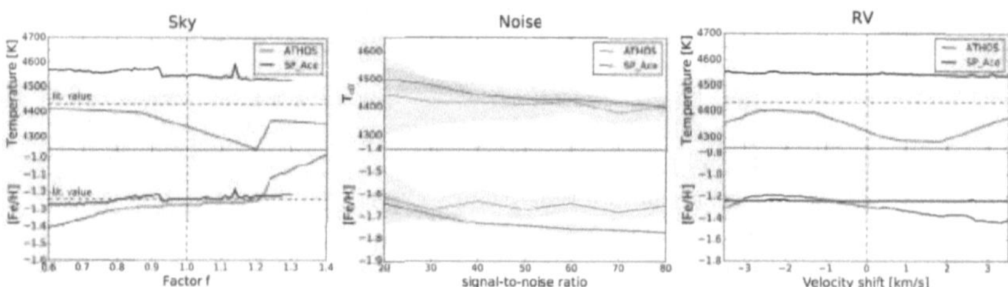

Figure 5.1.: Left panel: Impact of sky subtraction for 2MASS J00595966-3339319 on temperature and metallicity. The dashed line shows the literature values of Skúladóttir et al. (2015). Middle panel: Impact of noise on the stellar parameter. Here we used the star HD26297 as benchmark star. The band indicates a one sigma environment as we repeated the analysis 20 times. The signal-to-noise ratio is given per resolution element. Right panel: Uncertainty due to different radial velocity shifts. This plot was taken from Reichert et al. (2020).

5.2.1. Effective temperature

Since red giant stars in nearby dSph galaxies are relatively faint ($V \gtrsim 15\,\mathrm{mag}$), most of the spectra have relatively low S/N (as shown in Table 5.1). Therefore, the majority of literature studies relies on photometrically determined temperatures, while high-resolution or high S/N studies tend to rely on spectra, where the effective temperatures are determined from excitation equilibrium.

Depending on metallicity, we use ATHOS and SP_Ace, which derive effective temperatures differently. ATHOS relies on flux ratios, whereas SP_Ace fits an entire spectral region. The derived temperatures from each code are compared to the literature values in Table 5.2. Furthermore, we determined photometric temperatures based on the calibration of Alonso et al. (1999, 2001). We transformed the colors according to Bessell (1979) as well as Alonso et al. (1998) and applied a dereddening according to Schlafly and Finkbeiner (2011) using values from the IRSA Dust Database[5]. In contrast to most of the literature studies, we use an individual extinction for every star rather than relying on one value per galaxy. However, most galaxies are located far from the Galactic plane (Fig. 2.12) and the reddening correction is therefore small. This results in a negligible effect on the temperature (for $T_{\mathrm{eff}}(B - V)$ in, e.g., Sculptor the difference is $\sim 5\,\mathrm{K}$) caused by the variation in the local vs. global dereddening. The photometry was taken from various literature studies based on the SIMBAD Database (Wenger et al. 2000). The spread caused by using different methods is $\sim 150\,\mathrm{K}$, which is the average error we expect on the temperature. Photometric temperatures are on average lower than derived spectroscopic temperatures (see Table 5.2).

---

[5]https://irsa.ipac.caltech.edu/applications/DUST/, accessed on 13 March 2019

ATHOS has a higher spread compared to SP_Ace due to a higher sensitivity to velocity shifts, sky subtraction, and S/N, which may arise owing to the reduced number of covered flux ratios. However, the stars may also be out of the trained parameter range (cf. Fig. 2 of Hanke et al. 2018, and Table 5.3). Table 5.2 gives the average residual temperature $\langle \Delta T_{\mathrm{eff}} \rangle$ of SP_Ace and ATHOS compared to the literature values. In addition, we list the standard deviation of this residual, $\sigma(\Delta T_{\mathrm{eff}})$. Some stars show a large discrepancy between our methods caused by noisy spectra around the Balmer lines or strong stellar winds as in the Sculptor star 2MASS J00592830-3342073 (Fig. 5.2). The presence of stellar winds can cause emission lines around the Balmer lines (see, e.g., Olson and Ebbets 1981, Ebbets 1982, Haucke et al. 2018).

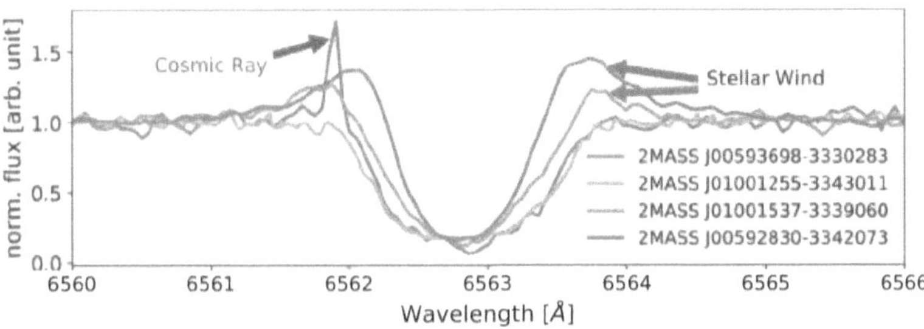

Figure 5.2.: Hα line for four different stars. There are clear features visible from cosmic rays and stellar winds visible.

Owing to the higher sensitivity to velocity shifts, reduced number of covered flux ratios, low S/N, and sky subtraction of ATHOS, we decided to adopt stellar parameters including effective temperatures determined by SP_Ace for all stars within its valid metallicity range respecting the boundary ([Fe/H] > -2.3). The temperatures of the more metal-poor stars were determined with ATHOS. The choice of using spectroscopically determined effective temperatures was made as the different photometric temperatures typically have an offset of $\sim 100\,\mathrm{K}$ (tested on 43 stars in Sculptor). A similar discrepancy between photometric temperatures of different colors was also noted by Letarte et al. (2010), Hansen et al. (2012), and Hill et al. (2019). In the study of Letarte et al. (2010), constant offsets of $\sim 100\,\mathrm{K}$ are applied to their adopted $T_{\mathrm{eff}}(V - \{J, H, K_s\})$ to account for systematics between the colors. In Hill et al. (2019), this discrepancy is on average $\sim 200\,\mathrm{K}$ for $T_{\mathrm{eff}}(V - I)$ compared to the photometric temperatures of other colors. In addition, our analysis revealed differences in temperature, depending on the photometric sources. Hill et al. (2019) rely on the photometry of Battaglia et al. (2008) for V and I and that of Babusiaux et al. (2005) for J and K_s, whereas Kirby et al. (2010) rely on the photometry of Westfall et al. (2006) for Sculptor. These various photometric calibrations can lead to differences of the order of $\sim 200\,\mathrm{K}$ for $T_{\mathrm{eff}}(V - I)$. In total, we determined $T_{\mathrm{eff}}(B - V)$, $T_{\mathrm{eff}}(V - K_s)$, and $T_{\mathrm{eff}}(V - I)$ for 145, 207, and 233 dSph galaxy stars, respectively. Not all colors are available for all stars, and we therefore decided to use spectroscopic temperatures to avoid these systematic offsets between colors and photometric sources. This increases the homogeneity of our sample, with the disadvantage of slightly larger uncertainties.

Table 5.2.: Average offset and spread of adopted effective temperatures compared to the literature values, where $\Delta T_{\mathrm{eff}} = T_{\mathrm{eff,lit}} - T_{\mathrm{eff,SP\_Ace/ATHOS}}$. "Method T_{lit}" indicates the method used in the literature to determine the effective temperature: photometrically (Phot), spectoscopically (Spec), or with fits over spectral regions (Fit).

Study	Galaxy	Method T_{lit}	$\langle\Delta T_{\mathrm{eff}}\rangle_{\mathrm{ATHOS}}$	$\sigma(\Delta T_{\mathrm{eff}})_{\mathrm{ATHOS}}$	# ATHOS	$\langle\Delta T_{\mathrm{eff}}\rangle_{\mathrm{SP\_Ace}}$	$\sigma(\Delta T_{\mathrm{eff}})_{\mathrm{SP\_Ace}}$	# SP_Ace
Letarte et al. (2010)	For	Phot	261	-	1	−99	51	72
Kirby et al. (2010)	For	Fit/Spec	-	-	-	−87	65	16
Kirby et al. (2013)	For	Fit/Spec	-	-	-	−217	73	20
Lemasle et al. (2014)	For	Phot	68	56	2	−235	66	36
Skúladóttir et al. (2015)	Scl	Phot	−57	163	7	−157	82	72
Hill et al. (2019)	Scl	Phot	−57	163	7	−158	79	71
Geisler et al. (2005)	Scl	Phot	-	-	-	−191	74	2
Simon et al. (2015)	Scl	Spec	3	-	1	-	-	-
Kirby et al. (2010)	Scl	Fit/Spec	−45	-	1	−100	87	6
Kirby et al. (2013)	Scl	Fit/Spec	−41	80	4	−100	226	44
Kirby et al. (2010)	Sex	Fit/Spec	-	-	-	−172	-	1
Kirby et al. (2013)	Sex	Fit/Spec	205	249	7	58	251	18
Shetrone et al. (2001)	Sex	Spec	-	-	-	−172	-	1
Sbordone et al. (2007)	Sgr	Phot	-	-	-	−96	114	12
Hansen et al. (2018)	Sgr	Spec	−89	367	2	−130	81	11
Bonifacio et al. (2004)	Sgr	Phot	-	-	-	26	66	9
Koch et al. (2008)	Car	Phot	-	-	-	−167	-	1
Norris et al. (2017)	Car	Phot	−50	-	1	−61	115	9
Fabrizio et al. (2012)	Car	Phot	−110	-	1	−78	119	17
Lemasle et al. (2012)	Car	Phot	174	341	4	−112	150	29
Ural et al. (2015)	UMi	Spec	129	-	1	−23	21	2
All studies	-	-	132	260	62	−119	137	462

Table 5.3.: Range of validity for different methods to derive stellar parameters.

Method	$T_{\mathrm{eff,min}}$ [K]	$T_{\mathrm{eff,max}}$ [K]	$\log g_{\min}$ [dex]	$\log g_{\max}$ [dex]	[Fe/H]$_{\min}$ [dex]	[Fe/H]$_{\max}$ [dex]
ATHOS	4000	6500	1.0	5.0	−4.5	0.3
SP_Ace	3600	7400	0.2	5.4	−2.4	0.4

5.2.2. Surface gravity

We determined the surface gravity photometrically, spectroscopically, and by using isochrones (Sect 4.3.2). In the following, we shortly summarize the used methods to derive the surface gravity.

The code SP_Ace (Boeche and Grebel 2016) creates a synthetic model and compares this to a large wavelength range of the spectrum (5212 − 6860 Å). The derived surface gravity is therefore based on ionization balance. The surface gravity is systematically offset by $\sim 0.3\,\mathrm{dex}$ compared to literature values. Similarly, ATHOS (Hanke et al. 2018) measures flux ratios of Fe I and Fe II lines and based on this, determines the surface gravity (Fig. 5.3).

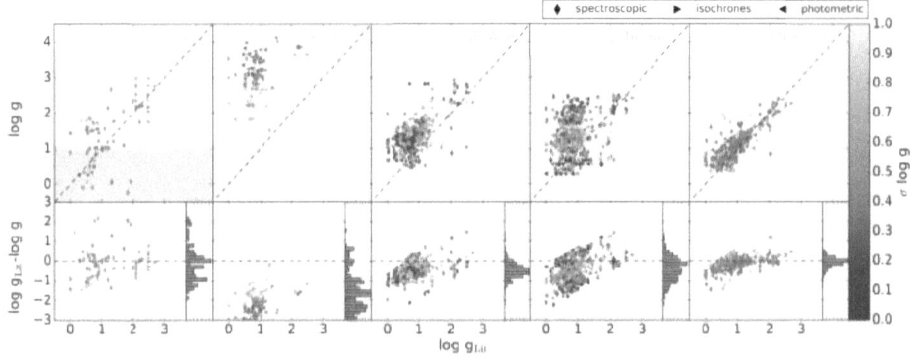

Figure 5.3.: Comparison between literature surface gravities and our calculated values. Upper panels show the surface gravity calculated using different methods versus the literature surface gravity. Symbols indicate the method that was used to derive the literature value and colors indicate the error of the here calculated value. Lower panels show the difference between the here derived values and literature values together with a histogram. Shaded areas indicate regions where ATHOS is out of the trained range. We adopt surface gravities shown in the upper right panel (distances).

We used Yale-Yonsei (YY) isochrones (Demarque et al. 2004) and employed a next neighbor interpolation in a grid of previously calculated isochrones. We applied a constant age of $10\,\mathrm{Gyr}$, an enhanced α-abundance $[\alpha/\mathrm{Fe}] = 0.3$, and assumed that our sample consists of giants only. The error was determined by applying an age error of $\Delta t_{\mathrm{Age}} = 1\,\mathrm{Gyr}$ including the errors from the derived temperatures and metallicities. Compared to literature values, the surface gravity spreads around $\sim 1\,\mathrm{dex}$, with the majority of the stars showing higher surface gravities.

Photometrically, we determined the surface gravities with two different distances. On one hand, the parallaxes of $\sim 1/3$ of our sample stars were recently presented by *Gaia* data release 2 (DR2, Gaia Collaboration 2018b). In fact, with Eq. (4.15), we can derive the surface gravity. However, the parallax has been measured only for a minority of our sample stars, because they are too faint. On the other hand, the distances of the dSph galaxies are well known by using variable stars as standard candles (Table 2.3). In combination with Eq. (4.14), the surface gravities can also be obtained (5th panel in Fig. 5.3). Here, we used bolometric magnitudes according to the calibration of Alonso et al. (1999). For stars where the $V-$magnitude is not known, we use *Gaia*'s broadband $G-$magnitude (Gaia Collaboration 2016) and calibrations of Andrae et al. (2018). We tested if this choice has a large offset for stars, where both magnitudes are known and found no fundamental difference (see Fig. 5.4).

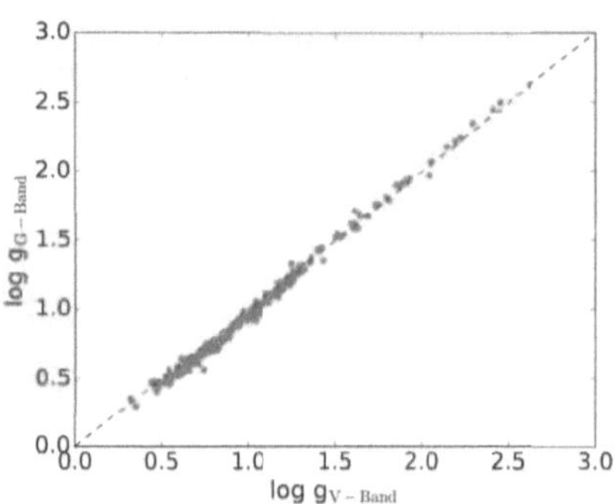

Figure 5.4.: Comparison of surface gravities determined with *Gaia*'s $G-$magnitude and the V-magnitude.

5.2.3. Metallicity

As for the temperature, we determined the metallicity ([Fe/H]) via SP_Ace and ATHOS. On average, ATHOS and Sp_Ace provide values that are approximately $0.1\,\mathrm{dex}$ more metal rich than the considered literature values, with a standard deviation of $0.38\,\mathrm{dex}$ and $0.28\,\mathrm{dex}$ for ATHOS and Sp_Ace, respectively. This systematic difference can be explained by higher temperatures compared to literature values. We analyzed the offset between ATHOS and Sp_Ace for stars in Sculptor, where both codes are operating within their boundaries. For these stars, both codes agree within a standard deviation of $\sigma = 0.15$ and the average offset is $[\mathrm{Fe/H}]_{\mathrm{ATHOS}} - [\mathrm{Fe/H}]_{\mathrm{Sp\_Ace}} = 0.09$ (Fig. 5.5). We want to stress, that the fairly large scatter is mainly driven by large errors in metallicity obtained with ATHOS ($\sim 0.3\,\mathrm{dex}$), which uses only a fraction of flux ratios for FLAMES/GIRAFFE spectra (11 out of 31 for most stars, see Table 5.4). For lower metallicities, most spectra are observed with UVES or HIRES and more flux ratios are used within ATHOS. The average offset of $0.09\,\mathrm{dex}$ is the trade-off in homogeneity that we have to consider when using these two different methods. However, this value is much lower than our error on the metallicity. We additionally note that ATHOS was trained on metallicities determined with MOOG and the equivalent width database of SP_Ace relies on MOOG as well. We therefore do not attempt to reanalyse iron abundances and take the value given by the corresponding code.

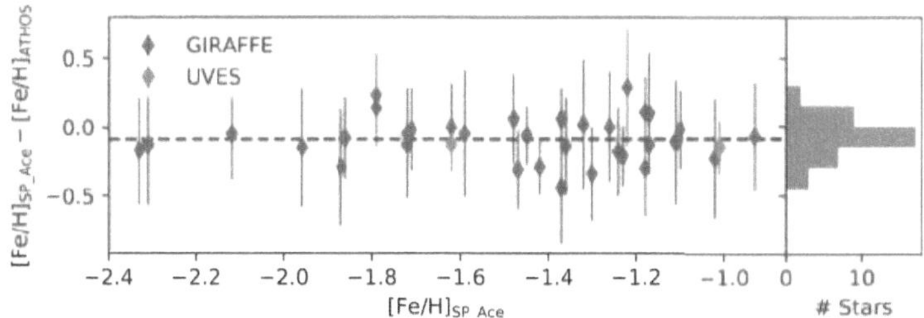

Figure 5.5.: Metallcity determined with SP_Ace and ATHOS for stars in Sculptor. Only stars for which ATHOS as well as SP_Ace operate within its range of validity are included.

Note that our metallicities are biased by not taking NLTE effects into account, because just like most other studies we rely on Fe I absorption lines as most Fe II lines are outside the covered wavelength range or they are too weak to detect them. According to Lind et al. (2012), a typical star in our sample ($T = 4500\,$K, $\log g = 1$, Fe/H $= -1.5$) would have a higher abundance/metallicity in NLTE by $\sim 0.1\,$dex. This correction may differ for varying stellar parameters and is in general stronger for lower metallicities. It can reach a maximum value of $0.4 - 0.5\,$dex adopting low surface gravities and metallicities (Lind et al. 2012, Amarsi et al. 2016).

Table 5.4.: Number of covered flux ratios of ATHOS for different setups and number of stars measured with the setups. HR denotes the high-resolution mode of FLAMES/GIRAFFE.

Setting	$\lambda_{\mathrm{min}}\,[\text{Å}]$	$\lambda_{\mathrm{max}}\,[\text{Å}]$	T	Fe	$\log g$
HR7A	4700	4974	5	2	1
HR10	5340	5620	0	11	1
HR13	6120	6400	0	0	0
HR14A	6390	6620	4	2	0
HR15	6610	6960	0	0	0
UVES	4800	6800	9	31	11

5.3. Testing the empirical metallicity relation for CEMP-stars

Now that we have derived the stellar parameter of 380 giant and super-giant stars, we take the opportunity to test the empirical relation to determine the metallicity of CEMP-stars, Eq. (4.19) and Eq. (4.21). Even when our sample includes only a few CEMP-stars, the stellar parameters are similar. For this analysis we took only a subsample of 300 stars. For 80 stars, no Ni and Cr was derived.

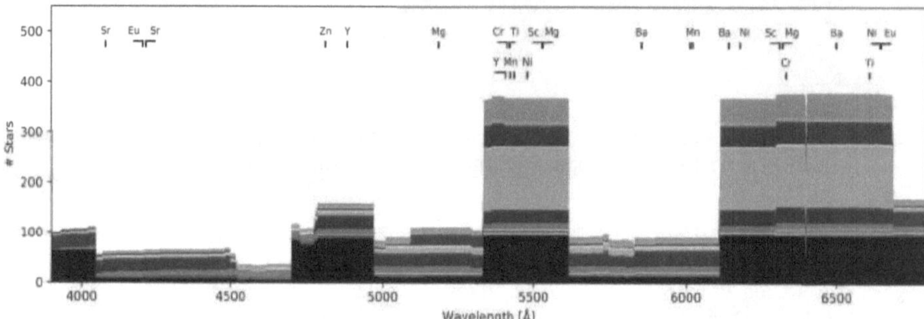

Figure 5.6.: Wavelength coverage and position of the individual spectral lines. Colors indicate different galaxies as in Fig. 2.12. The ionization state of each individual line is given in Table 5.5. This plot was taken from Reichert et al. (2020).

Most of our stars are observed with FLAMES/GIRAFFE only and the Cr lines around 5204 Å are not covered (see Fig. 5.6). Therefore, we measured Cr at 5409.77 Å and generalized this to the Cr lines at 5204 Å and 5208 Å by calculating the expected EWs from the abundances. We stress that such calculation requires knowledge of all stellar parameters and that Cr is extremely temperature sensitive. As a consequence, this will introduce additional uncertainties compared to a direct measurement of the EW. Furthermore, analogues to before, we had to mimic the Fe blends of the line (Fig. 4.14). The usage of the metallicity will reduce the error of our relation again. The large upturn in Fig. 5.7 of the $Cr_{5208.51}$ may be a saturation effect of the strong Cr line at the highest metallicities.

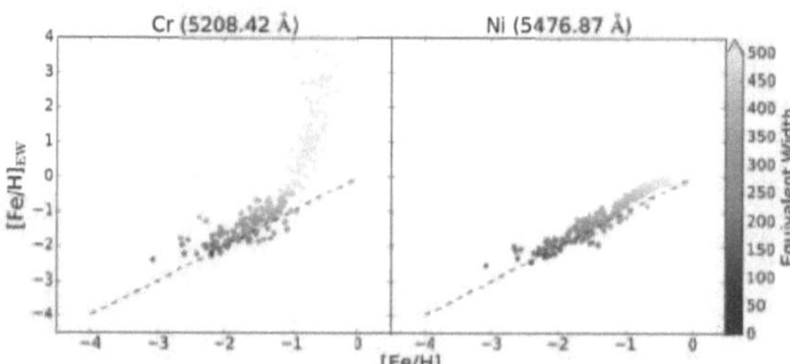

Figure 5.7.: Metallicities from SP_Ace and ATHOS compared to metallicities derived from EW relations (Cr left, Ni right) in a sample of 230 C-normal giants and supergiants with GIRAFFE spectra. This plot was taken from Singh et al. (2020).

For a smaller sample (59 stars) that have been measured with UVES, we are able to directly measure the EWs. However, due to the higher resolution, we had to add the Fe EW to the Ni line in order to mimic the situation in lower-resolution spectra. In total, we derive a range of validity that is given for the $Cr_{5208.51}$-relation within the boundaries of $50\,\mathrm{m\AA} < EW < 450\,\mathrm{m\AA}$, which translates to metallicities between $-2.8\,\mathrm{dex} \leq [Fe/H] \leq -0.8\,\mathrm{dex}$ (**Fig. 5.8**). For the $Ni_{5476.87}$-relation, the validity is given for $EW > 30\,\mathrm{m\AA}$, and metallicities $[Fe/H] > -3.2\,\mathrm{dex}$. Within these boundaries, we derived residual standard deviations of $\sigma_{5204.51} = 0.53\,\mathrm{dex}$, $\sigma_{5206.04} = 0.47\,\mathrm{dex}$, $\sigma_{5204.51+5208.42} = 0.44\,\mathrm{dex}$, $\sigma_{5208.42} = 0.39\,\mathrm{dex}$, and $\sigma_{5476.87} = 0.32\,\mathrm{dex}$ (**Fig. 5.8**). For the larger FLAMES/GIRAFFE sample (**Fig. 5.7**) we derive $\sigma_{5208.42} = 0.29$ and $\sigma_{5476.87} = 0.20$. Here, the $Cr_{5208.51}$-relation, the combined $Cr_{5204.51+5208.42}$-relation and the $Ni_{5476.87}$-relation provide the best metallicity trace. However, only Ni fulfills our goal and stays within $\pm0.2\,\mathrm{dex}$ uncertainty.

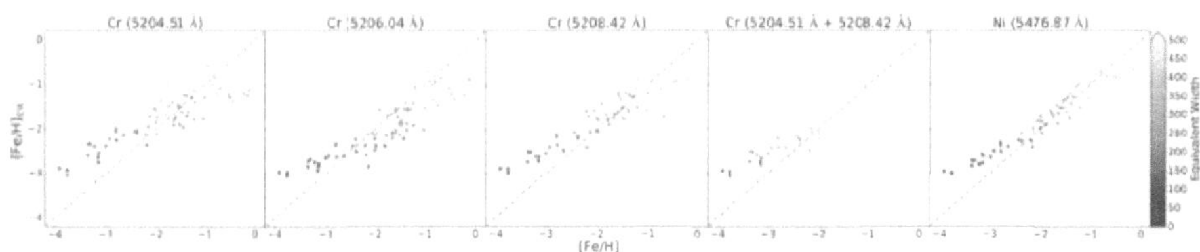

Figure 5.8.: Metallicities from Fe lines (SP_Ace and ATHOS) compared to metallicities derived from EW relations in a UVES sample of 59 C-normal giants and supergiants. The line strength is indicated by color. This plot was taken from Singh et al. (2020).

We succeeded in providing an empirical relation that is more accurate than common methods to derive stellar parameters of CEMP-stars. There exists the caveat, that giant and dwarf stars have to be treaded separately due to the very different line strength and line behavior. Here, we only showed a consistency test of giant stars, however Singh et al. (2020) also shows similar tests for dwarf stars. The derived empirical formulas will tremendously ease the metallicity determination of future surveys that encompass numerous CEMP-stars.

5.4. Abundance analysis

To derive stellar abundances we use the LTE spectral synthesis code MOOG (Sneden 1973a, version 2014) together with the aforementioned 1D Kurucz atmosphere models.

Using different spectral lines for the analysis can cause systematic uncertainties (e.g., due to uncertainties in atomic data), which makes it difficult to exclude biases. We try to minimize this systematic uncertainty by attempting to measure the same lines for all stars. This is not always possible, because the wavelength of the absorption line may not be covered by the instrument (Fig. 5.6). We therefore try to use lines that are covered by most of our sample and avoid lines that are only covered in a minority. In addition, different absorption lines may be either too weak or too strong, depending on the metallicity of the star. However, stars with similar stellar parameters should be similarly affected by systematic uncertainties.

Another source of uncertainty is the applied method to determine abundances, i.e., either by measuring equivalent widths or by running a spectral synthesis. The first method requires a technique to deblend

Table 5.5.: Used linelist of spectral absorption lines.

Element	Wavelength [Å]	Excitation potential [eV]	log(gf)	Weights	Lit.
Mg I	5183.60	2.700	-0.17	1.0	Pehlivan Rhodin et al. (2017)
Mg I	5528.48	4.330	-0.51	1.0	Pehlivan Rhodin et al. (2017)
Mg I	6318.72	5.104	-2.02	1.0	Pehlivan Rhodin et al. (2017)
Mg I	6319.24	5.104	-2.24	1.0	Pehlivan Rhodin et al. (2017)
Mg I	6319.50	5.104	-2.72	1.0	Pehlivan Rhodin et al. (2017)
Sc II	5526.80	1.767	0.13 (HFS)	1.0	Lawler and Dakin (1989)
Sc II	6309.90	1.496	-1.52 (HFS)	1.0	Lawler and Dakin (1989)
Ti II	5418.77	1.582	-2.13	1.0	Wood et al. (2013)
Ti II	6606.95	2.060	-2.79	1.0	Martin et al. (1988)
Cr I	5409.77	1.029	-0.67	1.0	Sobeck et al. (2007)
Cr I	6330.09	0.941	-2.92	1.0	Sobeck et al. (2007)
Mn I	5420.36	2.141	-1.46 (HFS)	1.0	Kurucz (2011)
Mn I	5432.55	0.000	-3.80 (HFS)	1.0	Kurucz (2011)
Mn I	6013.51	3.070	-0.35 (HFS)	1.0	Den Hartog et al. (2011)
Mn I	6021.82	3.073	-0.05 (HFS)	1.0	Den Hartog et al. (2011)
Ni I	5476.92	1.825	-0.89	0.3	Roederer and Lawler (2012)
Ni I	6176.82	4.085	-0.26	1.0	Wood et al. (2014)
Ni I	6177.25	1.825	-3.51	1.0	Kostyk (1982)
Ni I	6643.56	1.676	-2.30	0.5	Lennard et al. (1975)
Zn I	4810.54	4.075	-0.17	1.0	Biemont and Godefroid (1980)
Sr II	4077.71	0.000	0.16 (HFS)	1.0	Bergemann et al. (2012)
Sr II	4215.52	0.000	-0.16 (HFS)	1.0	Bergemann et al. (2012)
Y II	4883.68	1.084	0.07	1.0	Kurucz (2011)
Y II	5402.78	1.838	-0.51	1.0	Kurucz (2011)
Ba II	5853.69	0.604	-1.01 (HFS)	1.0	Gallagher et al. (2012)
Ba II	6141.73	0.704	-0.08 (HFS)	0.5	Gallagher et al. (2012)
Ba II	6496.90	0.604	-0.38 (HFS)	1.0	Gallagher et al. (2012)
Eu II	4205.05	0.000	0.21 (HFS)	1.0	Lawler and Dakin (1989)
Eu II	6645.21	1.379	0.12 (HFS)	1.0	Lawler and Dakin (1989)

the absorption lines, which can be uncertain and most studies avoid these blended lines. We therefore perform spectrum synthesis using MOOG and manually inspect all lines that may be blended. We claim a successful detection if the absorption line is found to be at the 2σ significance level, and the line furthermore has be covered by at least three pixels. We did not attempt to automatize this step. In fact, we measured equivalent widths for chromium and titanium as they had clean lines, all other elemental abundances were synthesized.

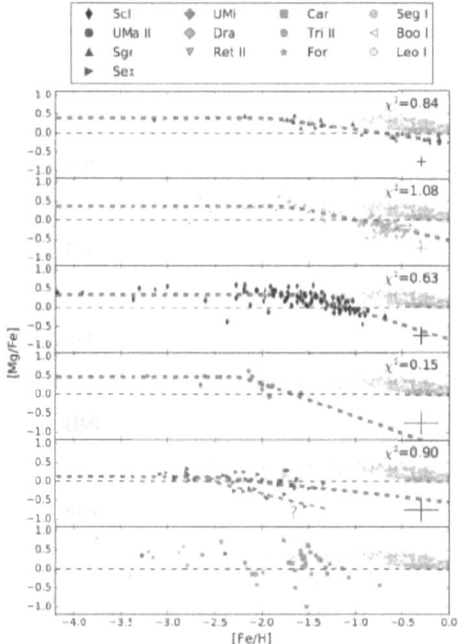

Figure 5.9.: Derived LTE abundances of [Mg/Fe] versus metallicities for individual dSph galaxies ordered by decreasing magnitude. The bottom panel shows the remaining dSph and UFD galaxies. Green triangles indicate stars of the globular cluster Terzan 7. Gray dots show MW stars of Gratton and Sneden (1988), Edvardsson et al. (1993), McWilliam et al. (1995), Ryan et al. (1996), Nissen et al. (1997), Hanson et al. (1998), Prochaska et al. (2000), Fulbright (2000), Stephens and Boesgaard (2002), Ivans et al. (2003), Bensby et al. (2003), and Reddy et al. (2003). All stars where the parallax differs significantly from the distance to the host galaxy are marked with red symbols (see Appendix D of Reichert et al. 2020). Stars with an error of [Mg/H] > 0.4 are indicated with open symbols. The reduced chi-square of the fit is given in the upper right corner of each panel. The median error for each galaxy is denoted in the lower right corner of each panel.

The following sections present the choice of adopted spectral lines as well as a brief discussion on NLTE corrections.

5.4.1. Magnesium

For magnesium, we determined the abundance from the Mg I feature at $\lambda = 5528.48\,\text{Å}$ (see Fig. 5.9). This line does not saturate at the highest metallicities and is detectable also in extremely metal-poor stars. Furthermore, we analyze the weakest line of the magnesium triplet at $\lambda = 5183.27\,\text{Å}$ and the region of the relatively weak Mg lines at $\lambda = 6318.71$, 6319.24, and $6319.50\,\text{Å}$ (see Table 5.5). We synthesized the convolved features centered around 6319.24 Å, and provide only one best-fitting Mg abundance for this blend. The atomic data of all Mg I lines are the recommended (theoretical) values from Pehlivan Rhodin et al. (2017). Neutral magnesium abundances are predominantly affected by temperature and their uncertainties are shown in Table 5.6. We also investigated the impact of this assumption by applying NLTE corrections from Bergemann et al. (2015) (accessed via the interface of Kovalev et al. 2018, see Fig. 5.14 and 5.13). In all galaxies, this leads to higher Mg abundances at low metallicities with corrections of $\sim 0.1\,\text{dex}$ (Fig. 5.14). We note that due to the limited coverage of the NLTE correction grid, all stars with $[\text{Fe/H}] < -2$ and simultaneously $\log g < 0.5$ are extrapolated from the grid[6].

5.4.2. Scandium

Sc II abundances are derived from the spectral line at $\lambda = 5526.80\,\text{Å}$. We also consider a weak line at $\lambda = 6309.90\,\text{Å}$. For both lines, we adopted atomic data and hyperfine splitting (HFS) from Lawler and Dakin (1989, see Table 5.5). The latter is a relatively weak line and only detectable in stars with $[\text{Fe/H}] \gtrsim -2$. It tends to yield lower ($\sim 0.2\,\text{dex}$) abundances compared to the bluer line. The standard deviation of both lines throughout our sample is $0.37\,\text{dex}$. Since both Sc lines are singly ionized, they are less affected by our LTE assumption, as confirmed by Zhang et al. (2008, 2014), who calculate almost negligible NLTE corrections for Sc II. The Sc LTE abundances can be found in Fig. 5.10.

5.4.3. Titanium

We measured the equivalent widths of Ti II at $\lambda = 5418.77\,\text{Å}$ and $6606.95\,\text{Å}$ (see Table 5.5). The Ti LTE abundances can be found in Fig. 5.10. We obtained similar results as Skúladóttir et al. (2017) and Hill et al. (2019), but lower values than Letarte et al. 2018 (see Fig. 5.10).

5.4.4. Chromium

We measured the equivalent widths of two Cr I lines depicted in Fig. 5.6. Most of the stellar spectra only include one line at $\lambda = 5409.77\,\text{Å}$, since the line at $\lambda = 6330.09\,\text{Å}$ is often too weak. The scatter in $[\text{Cr/Fe}]$ is significant and in addition driven by the temperature uncertainty (Fig. 5.10). Chromium can be highly affected by NLTE effects, especially when covering a large metallicity range (see, e.g., Mashonkina et al. 2017a). We apply NLTE corrections from Bergemann and Gehren (2008) to the Cr I lines (accessed via the interface of Kovalev et al. 2018). The correction, ΔNLTE, reaches extreme values between $0.77 \leq \Delta\text{NLTE} \leq 0.00$. Applying the corrections as illustrated in Fig. 5.14, we computed abundances that are slightly supersolar with average values of $\langle[\text{Cr/Fe}]_{\text{NLTE}}\rangle = 0.15$ in Fornax, Sagittarius and Sculptor, and $\langle[\text{Cr/Fe}]_{\text{NLTE}}\rangle = 0.0$ in Sextans (Fig. 5.13).

---

[6] the grid coverage can be accessed at http://nlte.mpia.de/grids.png

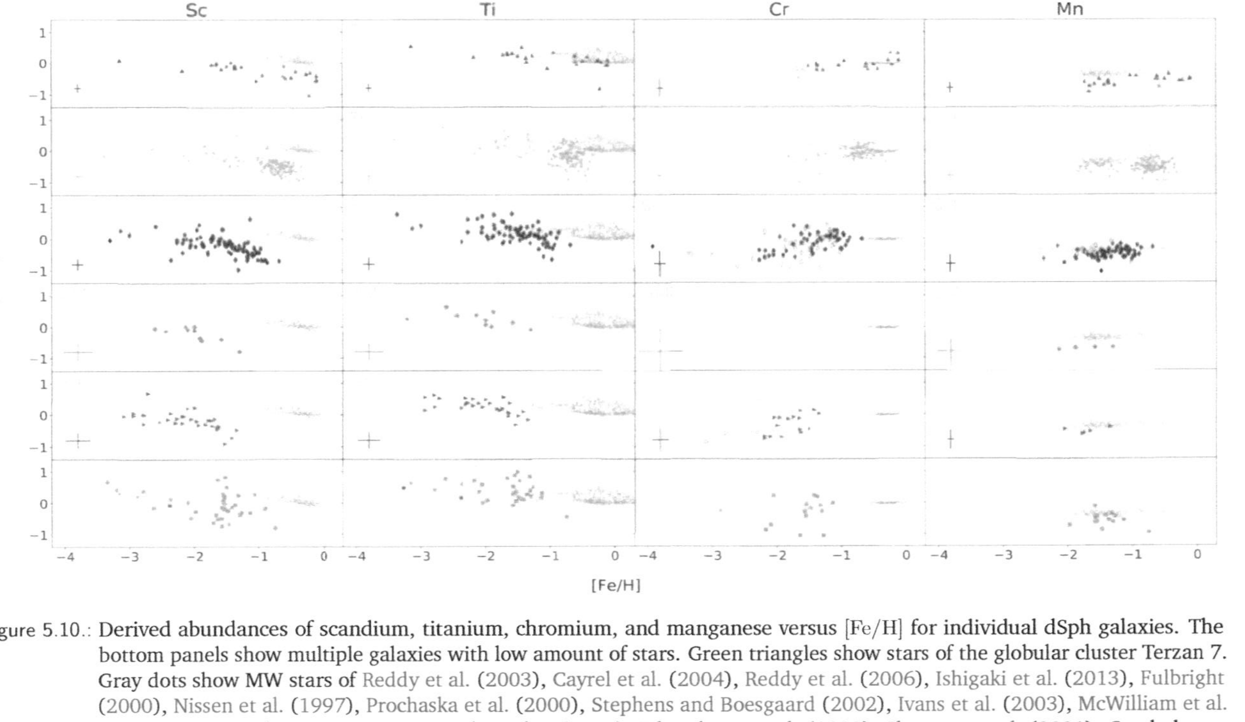

Figure 5.10.: Derived abundances of scandium, titanium, chromium, and manganese versus [Fe/H] for individual dSph galaxies. The bottom panels show multiple galaxies with low amount of stars. Green triangles show stars of the globular cluster Terzan 7. Gray dots show MW stars of Reddy et al. (2003), Cayrel et al. (2004), Reddy et al. (2006), Ishigaki et al. (2013), Fulbright (2000), Nissen et al. (1997), Prochaska et al. (2000), Stephens and Boesgaard (2002), Ivans et al. (2003), McWilliam et al. (1995), Ryan et al. (1996), Gratton and Sneden (1988), Edvardsson et al. (1993), Shetrone et al. (2001). Symbols are chosen as in Fig. 5.9, stars marked in red indicate stars with close distances according to the parallax (Appendix ??). This plot was taken from Reichert et al. (2020).

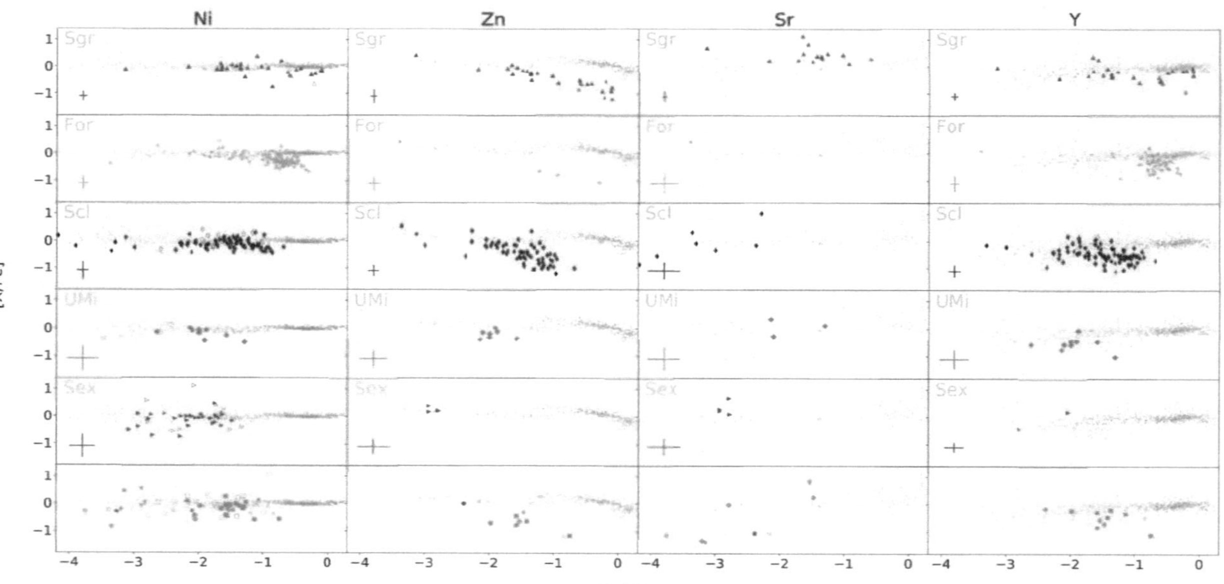

Figure 5.11.: Same as Fig. 5.10 for nickel, zinc, strontium, and yttrium versus [Fe/H] for individual dSph galaxies. Gray dots show MW stars of Reddy et al. (2003), Cayrel et al. (2004), Reddy et al. (2006), Ishigaki et al. (2013), Fulbright (2000), Nissen et al. (1997), Prochaska et al. (2000), Stephens and Boesgaard (2002), Ivans et al. (2003), McWilliam et al. (1995), Ryan et al. (1996), Gratton and Sneden (1988), Edvardsson et al. (1993), Shetrone et al. (2001). This plot was taken from Reichert et al. (2020).

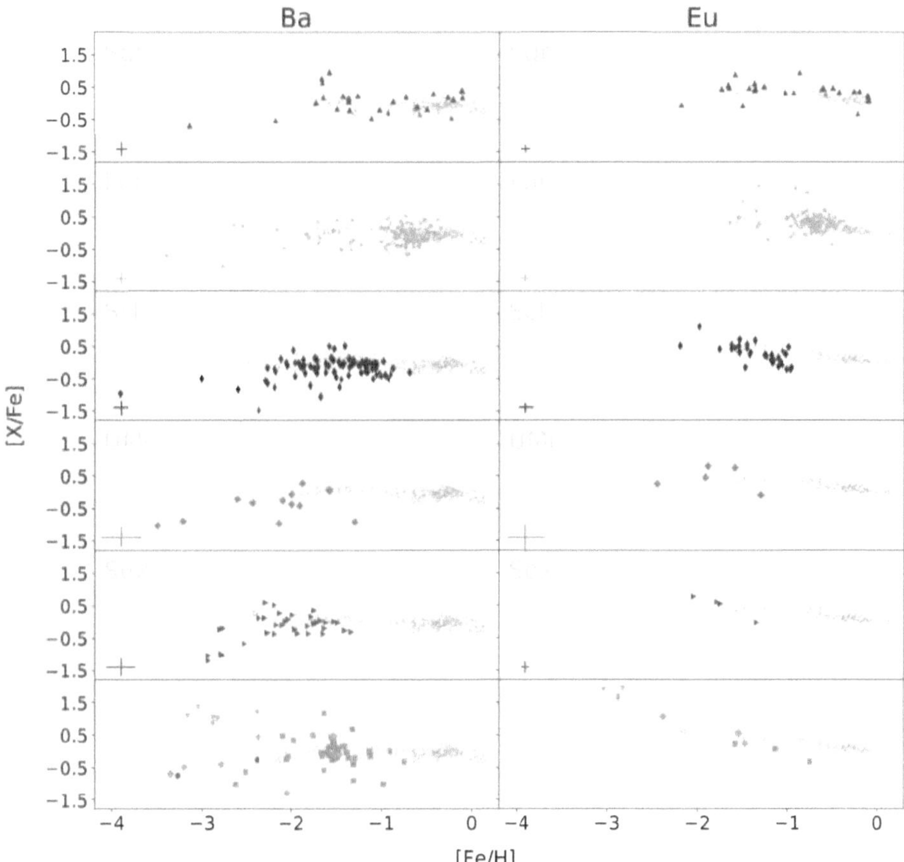

Figure 5.12.: Same as Fig. 5.10 for [Ba/Fe] and [Eu/Fe] versus [Fe/H]. This plot was taken from Reichert et al. (2020).

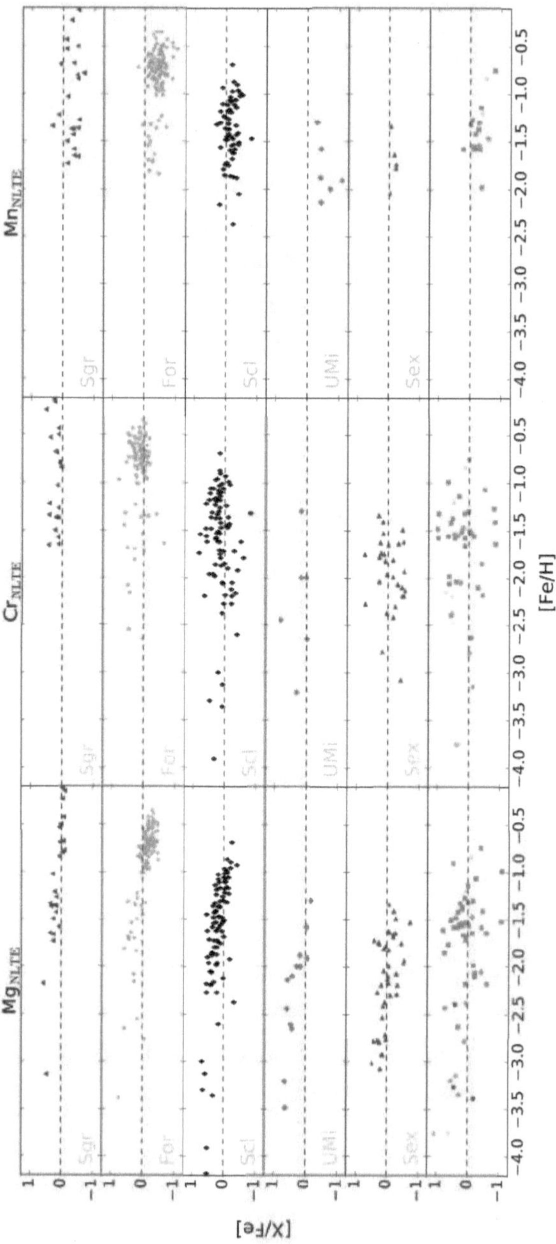

Figure 5.13.: NLTE corrected abundances of Mg, Cr, and Mn. Corrections are taken from Bergemann et al. (2015), Bergemann and Cescutti (2010), Bergemann and Gehren (2008) accessed via Kovalev et al. (2018). Symbols are chosen as in Fig. 5.9. This plot was taken from Reichert et al. (2020).

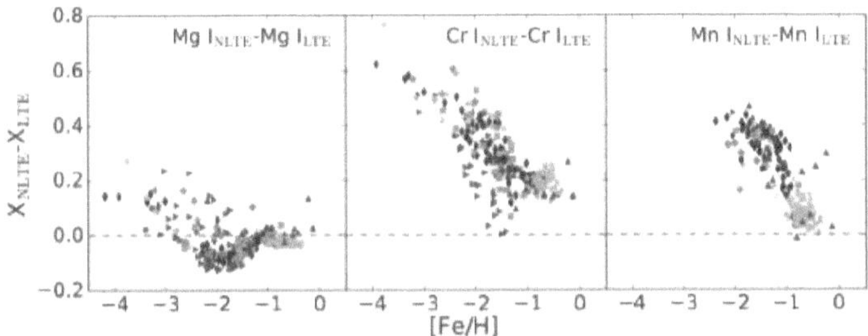

Figure 5.14.: NLTE corrections for Mg I, Cr I, and Mn I taken from Bergemann and Gehren (2008), Bergemann and Cescutti (2010), and Bergemann et al. (2015)(accessed via the interface of Kovalev et al. 2018). Symbols are chosen as in Fig. 5.9. This plot was taken from Reichert et al. (2020).

5.4.5. Manganese

We synthesized four manganese absorption lines (Table 5.5). For these we adopt the HFS from Kurucz (2011). North et al. (2012) reveal a large discrepancy between the lines at $\lambda = 5420.36$ Å and $\lambda = 5432.55$ Å for stars in Fornax that may be caused by LTE effects. The Mn LTE abundances can be found in Fig. 5.10. Manganese is highly affected by LTE assumptions. We therefore apply NLTE corrections from Bergemann and Gehren (2008) as shown in Fig. 5.14.

5.4.6. Nickel

We synthesized two weak Ni lines at $\lambda = 6176.82, 6177.25$ Å and two stronger lines at $\lambda = 5476.92, 6643.56$ Å. We adopt excitation potentials and oscillator strengths from Wood et al. (2013), Martin et al. (1988), Kostyk (1982) and Lennard et al. (1975). We assign different weights for the individual lines as indicated in Table 5.5. The Ni LTE abundances can be found in Fig. 5.11.

5.4.7. Zinc

We detected the zinc line at $\lambda = 4810.54$ Å in a sub-sample of our stars. This line is only covered by X-shooter, FLAMES/GIRAFFE HR7A, UVES, and the HIRES setup (see Fig. 5.6). We therefore measured zinc mainly for Sculptor and Sagittarius. The Zn LTE abundances are depicted in Fig. 5.11. We get the same trends as already presented in Skúladóttir et al. (2017) and Sbordone et al. (2007). For Sagittarius, we extend the current knowledge of zinc by also adding abundances from stars presented in Bonifacio et al. (2004), Monaco et al. (2005), and Hansen et al. (2018).

5.4.8. Strontium

We used the Sr II spectral line at $\lambda = 4077.71$ Å, which can be blended with La II and Dy II. For high metallicities, this line is saturated, resulting in a lower limit of the abundance. In addition, we included the slightly weaker line at $\lambda = 4215.52\,\text{Å}$. For both lines we assume HFS and isotopic shifts according to Bergemann et al. (2012). Unfortunately, this range is not covered in the FLAMES/GIRAFFE setups and we can therefore only extract Sr II in 38 stars. The assumption of LTE can affect the abundance of Sr, especially at $[\text{Fe/H}] < -3$ (Hansen et al. 2013, Mashonkina et al. 2017a). The LTE Sr abundances are illustrated in Fig. 5.11. In addition, our MOOG version does not include the effect of Rayleigh scattering (see, e.g., Sobeck et al. 2011). This effect is increasingly important for blue absorption lines in metal-poor stars and would lead to slightly higher Sr abundances if properly treated. We tested this for the bluest Sr II line and found that scattering on average increases the abundance by 0.03 dex for a typical metal-poor, cool giant ($T_{\text{eff}} = 4875\,\text{K}$, $\log g = 1.68$, $[\text{Fe/H}] = -1.73$). This value is much lower than our accuracy and precision. In most spectra we would not be able to distinguish this from the noise, but could only see this in our noise-free test synthesis. Strontium is the lightest n-capture element that we analyzed.

5.4.9. Yttrium

We analyzed two Y II lines, one strong line in the blue part of the spectra $\lambda = 4883.68\,\text{Å}$ and a weak redder line at $\lambda = 5402.78\,\text{Å}$. Similar to zinc, these lines are included in a minority of our sample owing to limited wavelength coverage. The $\log gf$ values for these lines are uncertain (see e.g., Ruchti et al. 2016). For the most metal-rich stars in Sculptor, we were able to measure both Y II lines. The comparison between the lines yields a constant offset of $\sim 0.2\,\text{dex}$, where the redder line results in higher abundances. The LTE yttrium abundances are presented in Fig. 5.11. So far there is no NLTE grid available, but we note that we analyze a spectral line of singly ionized Y that is formed in the lower photosphere where collisions are likely dominating.

5.4.10. Barium

We synthesized the three Ba II lines at $\lambda = 5853.69, 6141.73, 6496.90\,\text{Å}$ given in Table 5.5 and the resulting LTE abundances are shown in Fig. 5.12. We assign a lower weight to the line at $\lambda = 6141.73\,\text{Å}$, because it has been demonstrated to be heavily affected by the assumption of LTE (Korotin et al. 2015). We use HFS for all lines from Gallagher et al. (2012). According to Mashonkina et al. (2017a), we expect NLTE effects on the order of $0.3\,\text{dex}$ over the entire metallicity range (see also Mashonkina and Belyaev 2019). We derived much lower values for Ba in Fornax than previous studies. This difference is discussed in Appendix E of Reichert et al. (2020).

5.4.11. Europium

We measure one blue Eu II line at $\lambda = 4205.05\,\text{Å}$, which is present in high-resolution spectra only (UVES and HIRES) and a red Eu II line at $\lambda = 6645.21\,\text{Å}$. We are able to detect Eu II down to metallicities of $[\text{Fe/H}] \sim -2.7$. Both investigated europium lines do not yield a significant different abundance. Throughout all galaxies, Eu is slightly enhanced compared to our sun. NLTE corrections could increase the abundances by $\sim 0.15\,\text{dex}$ (Mashonkina et al. 2017a).

5.4.12. Error determination

We determined errors on the abundances by first converting our abundances to equivalent widths and we investigated how uncertainties in individual stellar parameters propagate into the abundances.

Table 5.6.: Abundance sensitivity for a representative virtual star with $T_{\mathrm{eff}} = 4400\,\mathrm{K}$, $\log g = 1$, $[\mathrm{Fe/H}] = -1.5$ and $\xi_t = 1.8\,\mathrm{km\,s^{-1}}$.

Element	ΔT_{eff} 140 K	$\Delta \log g$ 0.25 dex	$\Delta[\mathrm{M/H}]$ 0.2 dex	$\Delta \xi_t$ 0.15 km s^{-1}	σ_{tot} dex
$\Delta \log \epsilon(\mathrm{Mg})$	0.13	0.07	0.06	0.03	0.16
$\Delta \log \epsilon(\mathrm{Sc})$	0.02	0.11	0.07	0.01	0.13
$\Delta \log \epsilon(\mathrm{Ti})$	0.03	0.14	0.05	0.12	0.19
$\Delta \log \epsilon(\mathrm{Cr})$	0.30	0.05	0.05	0.10	0.32
$\Delta \log \epsilon(\mathrm{Mn})$	0.22	0.03	0.03	0.03	0.23
$\Delta \log \epsilon(\mathrm{Ni})$	0.19	0.01	0.01	0.06	0.20
$\Delta \log \epsilon(\mathrm{Zn})$	0.06	0.07	0.03	0.05	0.11
$\Delta \log \epsilon(\mathrm{Sr})$	0.04	0.04	0.01	0.16	0.17
$\Delta \log \epsilon(\mathrm{Y})$	0.01	0.09	0.05	0.02	0.11
$\Delta \log \epsilon(\mathrm{Ba})$	0.03	0.10	0.05	0.07	0.14
$\Delta \log \epsilon(\mathrm{Eu})$	0.08	0.05	0.06	0.07	0.13

We assumed the errors to be uncorrelated and computed a new model atmosphere with a new temperature that was offset by its adopted uncertainty. We then computed new abundances, which were compared to the adopted abundances. This resulted in $\sigma_{T_{\mathrm{eff}}}$. This was repeated for each of the stellar parameters and the differences were then added in quadrature resulting in a total uncertainty:

$$\sigma_{\mathrm{tot}} = \sqrt{\sigma_{T_{\mathrm{eff}}}^2 + \sigma_{[\mathrm{Fe/H}]}^2 + \sigma_{\log g}^2 + \sigma_{\xi_t}^2 + \sigma_{\mathrm{stat}}^2 + \sigma_{\mathrm{noise}}^2}. \tag{5.1}$$

Different elements have different sensitivity due to the stellar parameters (see Table 5.6). We note that estimating the error this way does not take into account the effects of possible blends. For elements where we measured only one absorption line, we weigh the error by the S/N per pixel at this line by assuming an additional uncertainty of

$$\sigma_{\mathrm{noise}} = \begin{cases} -0.01 \cdot \mathrm{S/N} + 0.4 & , -0.01 \cdot \mathrm{S/N} + 0.4 > 0 \\ 0 & , \text{else} \end{cases} \tag{5.2}$$

The total calculated error can be directly tested by comparing the derived abundances with literature values. The residual deviation of the abundances is illustrated in Fig. 5.15. The histogram reveals an approximate Gaussian shape that we fitted (dashed line). The FWHM of this Gaussian can also be predicted by taking the estimated errors of the abundances into account. We calculate the FWHM that we would expect from the errors as $\mathrm{FWHM} = 2.354 \cdot \sqrt{\sigma_{\mathrm{lit}}^2 + \sigma_{\mathrm{tot}}^2}$, where the value is illustrated as a horizontal line. Here σ_{lit}^2 is the average variance that the literature provides and σ_{tot}^2 is the average error on the abundance derived within this study. A smaller FWHM hints towards an underestimation of the error whereas a larger FWHM towards an overestimation or large systematic uncertainties between the studies. Nearly all elements show a perfect agreement of the theoretical and determined value of the FWHM. We therefore conclude that our derived errors are in a reasonable range.

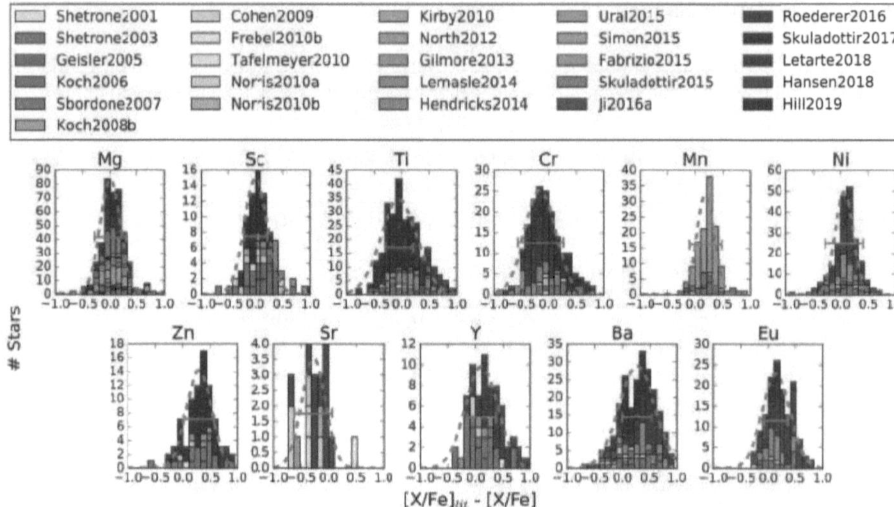

Figure 5.15.: Comparison of derived abundances with various literatures. The dashed line shows a Gaussian fit and the horizontal line indicates the FWHM one would expect due to the average error of the abundances. Literature values are taken from: Skúladóttir et al. (2015, 2017), Hendricks et al. (2014b), Cohen and Huang (2009), Hill et al. (2019), Letarte et al. (2018), Shetrone et al. (2003), Gilmore et al. (2013), Ural et al. (2015), Geisler et al. (2005), Koch et al. (2006, 2008), Roederer et al. (2016), Norris et al. (2010b,a), North et al. (2012), Sbordone et al. (2007), Simon et al. (2015), Ji et al. (2016a), Tafelmeyer et al. (2010), Kirby et al. (2010), Fabrizio et al. (2015), Shetrone et al. (2001), Lemasle et al. (2014), Frebel et al. (2010b), and Hansen et al. (2018). This plot was taken from Reichert et al. (2020).

5.5. Abundance trends

In the previous sections, we have presented a homogeneous set of stellar abundances across different galaxies. This set includes stars covering a large range of metallicities ($-4.18 \leq$ [Fe/H] ≤ -0.12). It gives us the unique possibility to study the characteristics of the individual systems. Each of the derived elements can trace individual processes and productions sites.

The α-elements are particular interesting for studying the star formation history (SFH) and the IMF (see e.g., Tinsley 1980, McWilliam 1997 and Sect. 2.5). We therefore investigate the evolution of α-elements in Sect. 5.5.1 to learn about the chemical history of these galaxies. Other elements like neutron-capture elements give insight to nuclear processes, such as the s-process hosted by AGB stars, and the r-process (e.g., in rare CC-SNe or NSM, see Sect. 2.3.1).

Figure 5.9 shows magnesium, which is created during hydrostatic burning (e.g., Truran and Heger 2003, Cayrel et al. 2004, Kobayashi et al. 2006, Nomoto et al. 2013). This element provides insight into the

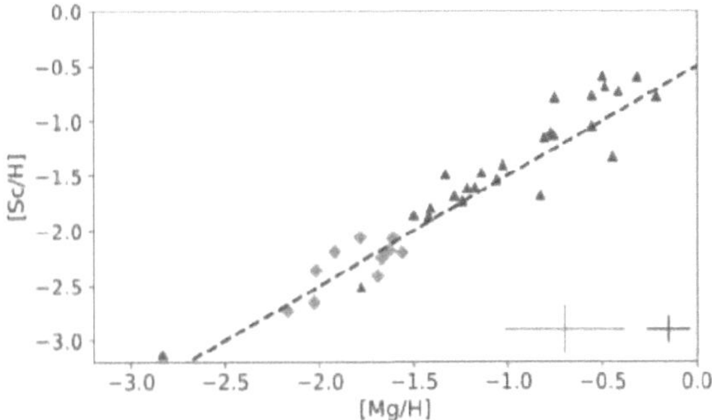

Figure 5.16.: [Sc/H] versus [Mg/H] for Sgr and UMi together with a 1 to 1 correlation. Symbols and colors are chosen as in Fig. 5.9.

IMF and star formation history. Sextans seems to have a lower [Mg/Fe] value at lower metallicities than other dwarf galaxies (cf., e.g., Fig. 2 of Aoki et al. 2009, Shetrone et al. 2001, Theler 2015, Theler et al. 2019). This deficiency of Mg at low metallicities is less striking when including NLTE-corrections (Fig. 5.13) and may therefore be a consequence of LTE assumptions. Another possible explanation might be a poor stochastical sampling of the IMF for small systems (François et al. 2016, Applebaum et al. 2018), resulting in lower [α/Fe] (Tolstoy et al. 2003, Carigi and Hernandez 2008).

Titanium on the other hand traces explosive burning, and forms predominantly in CC-SNe (e.g., Cayrel et al. 2004, Nomoto et al. 2013). We observe a similar evolution for Ti as for the MW with the exception of the position of the α-knee, which will be discussed in Sect. 5.5.1. In Fig. 5.10 we see a clear α-knee in both Sc and Ti. The abundance trend of scandium shows a good agreement with the one of magnesium (Fig. 5.16).

The Fe-peak elements, chromium, manganese, and nickel are produced by a mixture of type Ia SNe as well as (standard) CC-SNe in our sun (e.g., Bergemann and Cescutti 2010, North et al. 2012, Nomoto et al. 2013, Kirby et al. 2019). The heavier element zinc shows a decrease in Sagittarius and Sculptor at approximately the same position as magnesium. This decrease in combination with a low abundance scatter, points towards (standard) CC-SNe as main production channel (Fig. 5.11). Both chromium and manganese show decreasing trends with decreasing metallicities in LTE (Fig. 5.10). However, when correcting for LTE effects, both elements show flat trends (Fig. 5.13). Similarly, nickel shows a flat trend with [Fe/H] (Fig. 5.11) and only Fornax and possibly Sculptor show a large amount of clumping at higher [Fe/H], which would indicate a deviation from this flat trend. We note that compared to literature, this conclusion is different. In Hill et al. (2019) and similarly Kirby et al. (2019), a clear decreasing trend in nickel is visible.

We derive abundances of the lighter n-capture elements strontium and yttrium (see Fig. 5.11). According to Sneden et al. (2003) and Bisterzo et al. (2014), strontium is mainly produced by the s-process in our sun. However at low metallicities, it has also contributions from other processes such as the light element primary process (LEPP; Travaglio et al. 2004, Montes et al. 2007).

The heavy n-capture element barium is formed by the s-process (85% according to Bisterzo et al. 2014) in the solar system. This makes it an excellent tracer of the s-process. On the other hand, in the solar system europium is to 94 % made by the r-process (Bisterzo et al. 2014), making it our best r-process tracer. In Fig. 5.12 we see the chemical evolution of these two heavy elements. Most dSphs show a large star-to-star scatter and a tendency of decreasing heavy element to iron ratios below [Fe/H]$= -2$. Further trends will be discussed in Sect. 5.5.2.

5.5.1. The IMF and the alpha knee

To study the alpha knee we focus on two main production sites, namely CC-SNe forming the α-elements (i.e., Mg, Si, Ca, Ti) and type Ia SNe creating in large amounts the iron-peak elements (i.e., Fe, Ni, Zn). The former results from massive stars and can occur early; the latter includes white dwarfs in a binary system and appears later in the galactic history. Early on, the chemical enrichment is dominated by CC-SNe whose yields have high ratios of magnesium versus iron ($[\alpha/\mathrm{Fe}]_{\mathrm{CC-SNe}}$), while type Ia SNe yield lower ratios of magnesium versus iron later on (see Fig. 5.9). Observationally, the onset of type Ia SNe is seen as a decrease in α to iron, and it is known as the 'α-knee' ($[\mathrm{Fe/H}]_{\alpha,\mathrm{knee}}$). The location of this knee is the most striking difference between the investigated dwarf galaxies. To quantitatively describe the position, we assume a toy model with a plateau for low metallicities and a linear decrease after $[\mathrm{Fe/H}]_{\alpha,\mathrm{knee}}$ (for a similar approach see, e.g., Cohen and Huang 2009, Vargas et al. 2013, de Boer et al. 2014, Hendricks et al. 2014a, Kirby et al. 2019).

First, we determined the position of the α-knee for magnesium, scandium, and titanium individually (left panel of Fig. 5.17). We use Mg, Sc, and Ti, because they show a decrease at the same metallicity, even if they are not all typical α-elements. The α-knees were fit by an ordinary least squares fit.

In a next step, we calculated the error weighted mean and standard deviation of the knee position shown in Table 5.7. Hendricks et al. (2014a) found a much lower value of the α-knee in Fornax ($[\mathrm{Fe/H}]_{\mathrm{Mg,knee}} = -1.88$ and $[\mathrm{Fe/H}]_{\alpha,\mathrm{knee}} = -2.08$), based on different elements (Mg, Si, and Ti). The higher values of our work are related to high scandium values, whereas low values in Hendricks et al. (2014a) are mainly driven by silicon ($[\mathrm{Fe/H}]_{\mathrm{Si,knee}} = -2.49$). Chemical evolution models that successfully reproduce the metallicity distribution function in Fornax predict a knee at $[\mathrm{Fe/H}]_{\alpha,\mathrm{knee}} \approx -1.4$ (Kirby et al. 2011, Hendricks et al. 2014a). This is close to our values, however, Kirby et al. (2011) note that the applied chemical evolution model does not reflect the complete complex behavior of Fornax and the agreement between the model and our value may be a coincidence (see, e.g., Lanfranchi and Matteucci 2003, 2004, 2010 for other chemical evolution models of dSph galaxies or Cohen and Huang 2009, 2010, de Boer et al. 2012, Starkenburg et al. 2013, Kirby et al. 2019 for other observational derivations of the location of the knee).

Previous studies (e.g., Tinsley 1980, Matteucci and Brocato 1990, Gilmore and Wyse 1991, Venn et al. 2004, Tolstoy et al. 2009, Hendricks et al. 2014a) proposed a relation between the mass (and therefore the luminosity) of a galaxy and the position of the α-knee. In the following, we will use the luminosity of a galaxy as a proxy for its mass. We note, however, that for tidally disrupted galaxies, this may not be a good substitute (e.g., Tolstoy et al. 2009, McConnachie 2012). Nevertheless, a large deviation from the mass and luminosity relation should only apply for a strong tidal disruption, because the galaxies dark matter is thought to get disrupted before the stars and the impact on the luminosity may therefore be minor (see e.g., Smith et al. 2016). More massive systems are expected to keep their metals more efficiently, leading in most cases to the appearance of the α-knee at higher metallicities due to an extended SFH. By assuming a simple linear dependency of these quantities, we get a rough estimate of the knee

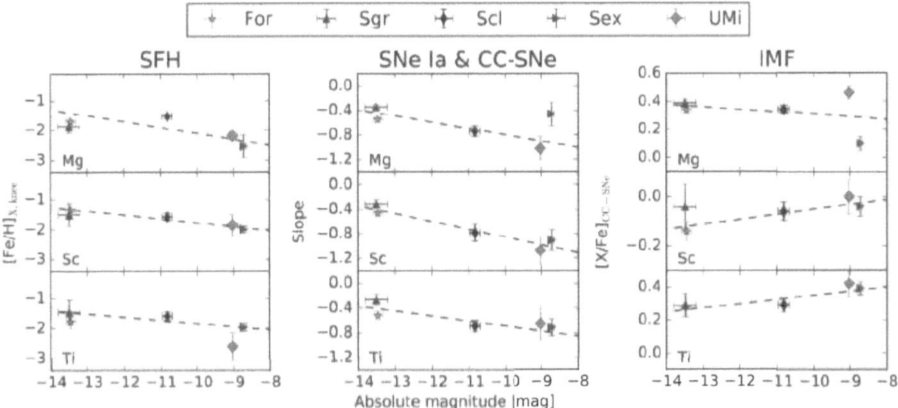

Figure 5.17.: Left panel: Measured position of the α-knee in five classical dSphs as a function of metallicity and absolute V magnitude. Middle panel: Slope of the decrease of [X/Fe] over [Fe/H] at metallicities higher than the locus of the α-knee. Right panel: Constant value of the plateau in [α/Fe] for different α-elements in the individual systems. Symbols are the same as in Fig. 5.9. This plot was taken from Reichert et al. (2020).

(in Fornax, Sagittarius, Sextans, and Ursa Minor, see Table 5.7):

$$[\text{Fe/H}]_{\alpha,\text{knee,estimate}} \approx (-0.10 \pm 0.03) \cdot M_V + (-3.00 \pm 0.25). \tag{5.3}$$

Here we included x- as well as y- errors in the fit adopting an orthogonal distance regression (ODR)[7]. We note that the fit contains large uncertainties due to the limited amount of data points. Nevertheless, this relation also fits reasonably well for the MW at $M_V = -20.8\,\text{mag}$ (Karachentsev et al. 2004, see Fig. 5.18). In addition to the knee position, the slope of each [α/Fe] also depends on the galaxy mass (middle panel of Fig. 5.17).

Table 5.7.: Approximate position of the α-knee and absolute magnitudes taken from McConnachie (2012) for Sagittarius and from Muñoz et al. (2018) for the other four dSphs.

Galaxy	$[\text{Fe/H}]_{\text{Mg,knee}}$	$[\text{Fe/H}]_{\text{Sc,knee}}$	$[\text{Fe/H}]_{\text{Ti,knee}}$	$[\text{Fe/H}]_{\alpha,\text{knee}}$	Absolute magnitude [mag]
Sgr	-1.87 ± 0.21	-1.48 ± 0.38	-1.47 ± 0.42	-1.67 ± 0.19	-13.50 ± 0.30
For	-1.71 ± 0.13	-1.35 ± 0.15	-1.77 ± 0.26	-1.59 ± 0.18	-13.46 ± 0.14
Scl	-1.52 ± 0.07	-1.57 ± 0.12	-1.59 ± 0.13	-1.55 ± 0.03	-10.82 ± 0.14
UMi	-2.18 ± 0.11	-1.84 ± 0.35	-2.61 ± 0.47	-2.18 ± 0.23	-9.03 ± 0.05
Sex	-2.53 ± 0.39	-1.99 ± 0.11	-1.96 ± 0.13	-2.05 ± 0.19	-8.72 ± 0.06

Simplified, this linear trend can be explained with a combination of the SFH and the relative amount of CC-SNe and type Ia SNe yields retained within each system. Imagine a smaller (absolute) amount of

---

[7]via the python package scipy.odr

Table 5.8.: Estimates of the α−knee with absolute magnitudes taken from Muñoz et al. (2018).

Galaxy	M_V [mag]	$\mathrm{[Fe/H]}_{\alpha,\mathrm{knee,estimated}}$
Leo I	-11.78 ± 0.28	-1.8 ± 0.1
Leo II	-9.74 ± 0.04	-2.0 ± 0.1
Car	-9.43 ± 0.05	-2.0 ± 0.1
Dra	-8.71 ± 0.05	-2.1 ± 0.1
Boo I	-6.02 ± 0.25	-2.4 ± 0.1
Her	-5.83 ± 0.17	-2.4 ± 0.1
Comber	-4.38 ± 0.25	-2.6 ± 0.2
UMa II	-4.25 ± 0.26	-2.6 ± 0.2
Ret II	-3.88 ± 0.38	-2.6 ± 0.2
Seg I	-1.3 ± 0.73	-2.9 ± 0.2

Fe in less massive systems due to an earlier halt of star formation, a lower gravitational potential, and the implied lower escape velocities. A single type Ia SN event will therefore lower the α/Fe-ratio by a larger factor. In addition, the total number of CC-SNe to type Ia SNe that enter the chemical evolution of observed stars may depend on the mass of the galaxy. The slopes of the decreasing trends of the elements depend on the rate of CC-SNe, the rate of type Ia SNe, the yields, and the SFHs. We note that this is a simplified view on the evolution of elements, as there are outliers from the here presented mass relation such as Carina, which may have undergone a merger event.

The Sextans dSph galaxy seems to show an indication of having two knees when looking at the magnesium abundances (dashed line in Fig. 5.9, cf. Fig. 3.6 of Theler 2015 and Theler et al. 2019). For more information of the two of our most metal-poor stars in Sextans, we refer the reader to Lucchesi et al. 2020). In order to test[8] whether a model with two knees should be favored over a model with one, we apply the Akaike information criterion (AIC, Akaike 1974) and the Bayesian information criterion (BIC, Schwarz 1978). These criteria take the goodness of the fit as well as the amount of parameters into account. For Sextans, the BIC favors a model with one cluster only, whereas the AIC favors a model with two clusters. However, the difference of the two models is negligible. From the current data it is not possible to get a clear answer or evidence, but future observations may provide better statistics to improve this. In addition, we are aware that titanium does not show a signature of a second knee. Nevertheless, we briefly discuss the implication of this finding.

One knee lies around $\mathrm{[Fe/H]}_{\mathrm{Mg,knee}} = -2.5$ and one around $\mathrm{[Fe/H]}_{\mathrm{Mg,knee}} = -2.0$. Cicuéndez and Battaglia (2018) found that Sextans may have undergone a galaxy accretion/merger event. They note that simulations (e.g., Benítez-Llambay et al. 2016) indicate that older stars (early population) may assemble in mergers, whereas the younger ones are formed in-situ. Following this idea, the knee at $\mathrm{[Fe/H]}_{\mathrm{Mg,knee}} = -2.0$ may therefore be consistent with the similar luminous Ursa Minor that has not undergone a merger event (Fig. 5.18), whereas the more metal-poor knee may be formed before the accretion. According to Eq. 5.3, a knee at $\mathrm{[Fe/H]}_{\mathrm{Mg,knee}} \sim -2.5$ would translate to $M_V \sim -5.2\,\mathrm{mag}$ and therefore $M_* \sim 10^4 M_\odot$. This agrees with masses of globular clusters or UFD galaxies comparable to Ursa Major I, which was previously pointed out to show remarkable similarities to Sextans (Willman et al. 2005). The lack of metals before the onset of type Ia SNe in Sextans could be also explained by a short main episode of star formation. Bettinelli et al. (2018) found that the main episode of star formation

---

[8]Using a Gaussian mixture clustering algorithm via the python package scikit-learn, Pedregosa et al. 2011

lasted only $\sim 0.6\,\mathrm{Gyr}$ and ended $\sim 12.9\,\mathrm{Gyr}$ ago, which supports the hypothesis that Sextans quickly drained most of its gas reservoir.

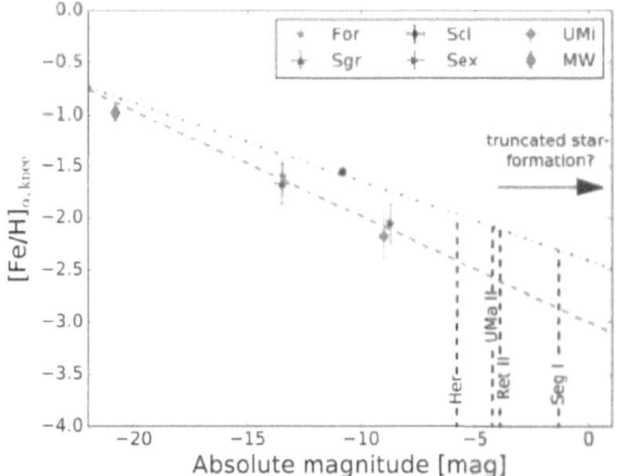

Figure 5.18.: Metallicity of the α-knee versus the absolute magnitudes of the parent dSph galaxy and the MW. The lines indicate a fit including and excluding Sculptor. This plot was taken from Reichert et al. (2020).

The IMF can be investigated by the value of the plateau of $[\alpha/\mathrm{Fe}]$ at low metallicities (e.g., Tinsley 1980, McWilliam 1997). The yields of more massive stars are predicted to have higher $[\alpha/\mathrm{Fe}]$ (e.g., Tsujimoto et al. 1995, Woosley and Weaver 1995, Kobayashi et al. 2006). The value of this plateau is shown in the right panel of Fig. 5.17. A linear fit of the plateau value results in slopes of -0.02 ± 0.03, 0.02 ± 0.01, and 0.02 ± 0.01 for Mg, Sc, and Ti, respectively. These almost negligible slopes indicate that there is no dependence of the IMF on the absolute magnitude of the system. Finally, we extrapolate the knee position (see Table 5.8) for small systems to test if the relation breaks down. We expect this relation to break down at the lowest galaxy masses, if they stopped forming stars before the onset of type Ia SNe (see e.g., Weisz et al. 2019), shown by a grey band in Fig. 5.18. Vargas et al. (2013) demonstrated with medium resolution DEIMOS spectra that this is the case for Ursa Major II and Segue I, but most UFDs are able to form stars long enough to maintain the signatures of type Ia SNe.

For small galaxies (low stellar masses), with insufficient data to directly fit a knee position, we can extrapolate the relation given in Eq. 5.3. Figure 5.19 indicates the knee position estimates for Leo I, Bootes I, Reticulum II, and Ursa Major II. In addition, we added literature values for the UFD galaxy Reticulum II (Ji et al. 2016b). The star [KCB2015] Reti 15 has an unusual low ratio of $[\mathrm{Sc}/\mathrm{Fe}]$. Only a few stars at low metallicities show such a deficiency in Sc (Casey and Schlaufman 2015). Our relation strengthens the hypothesis that this star (although low in metallicity) was formed in a pocket of the ISM where type Ia SNe already contributed (see also, e.g., Ji et al. 2016b). This is supported by the low ratio of $[\mathrm{Mg}/\mathrm{Fe}]$, but stands in contradiction to the high ratio of $[\mathrm{Ti}/\mathrm{Fe}]$.

The overall good agreement between all galaxies suggests that SFH, and the fraction of type Ia-, and CC-SNe yields are strongly dependent on the absolute magnitude (and therefore the stellar mass) of the galaxy. Having different SFHs and varying fractions of dark matter (for Sculptor, see, e.g., Massari et al. 2018) in these systems such a strong relation (Eq. 5.3 and depicted in Fig. 5.17) is not obvious.

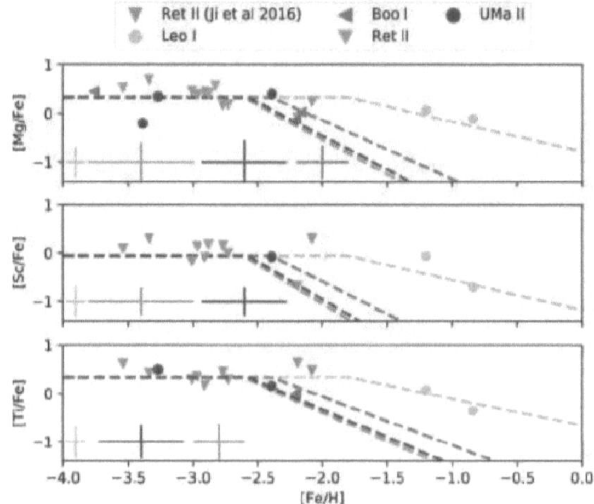

Figure 5.19.: Estimate of [Mg/Fe], [Sc/Fe], and [Ti/Fe] for Reticulum II, Leo I, Bootes I and Ursa Major II. The median of all errors for each galaxy is shown in the lower left of every panel. This plot was taken from Reichert et al. (2020).

5.5.2. Neutron-capture elements

Here we start by determining when the s-process first contributes to the chemical enrichment and later study possible hosts of the r-process in dSph systems. Furthermore, we discuss whether the stellar mass of the galaxies has an impact on the enrichment in heavy elements.

5.5.3. The s-process

The s-process is mainly hosted in AGB stars, but for the r-process there are multiple possible pathways. Since the gravitational wave detection of a binary neutron star merger GW170817 (LIGO Scientific Collaboration and Virgo Collaboration 2017) and the optical counterpart AT 2017gfo, these events have shown to be able to produce heavy elements, indicated by the transient (see e.g., Cowperthwaite 2017, Kasliwal et al. 2017, Chornock et al. 2017, Drout et al. 2017, Shappee et al. 2017, Pian et al. 2017, Waxman et al. 2018, Arcavi 2018, Watson et al. 2019). In the following, we want to recap arguments against and for NSMs being the dominant production site for the r-process elements in various dSph galaxies or whether an additional source is necessary (e.g., rare types of CC-SNe) to explain the abundance trends that were derived within this study.

In order to trace contributions from various nuclear formation processes we search for abundance correlations. An indication of two elements being formed by the same process can be obtained if the element abundance ratio [A/B] versus [B/H] shows a flat trend, indicative of the two elements, A and B, growing at the same rate as a function of the element B (Hansen et al. 2012). Alternatively, a very clean trace can be obtained by simply plotting the absolute abundances of A versus B where a 1:1 ratio (with low scatter) shows a correlation and process co-production (Hansen et al. 2014a). We stress that these two elements do not necessarily have to form in exactly the same event, but with the same time delay. A delayed production of A will result in a more positive slope of the absolute abundances of A versus B (Duggan et al. 2018, Skúladóttir et al. 2019).

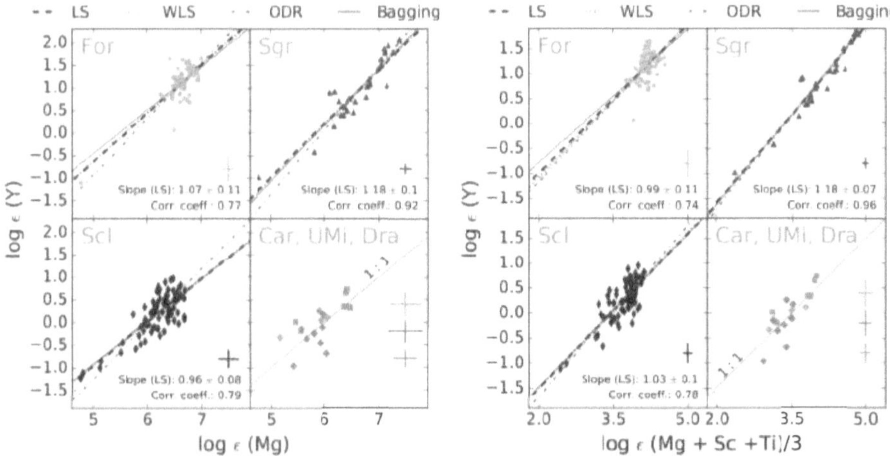

Figure 5.20.: Left panels: Absolute abundance of yttrium versus absolute abundance of magnesium together with a linear fit (Reichert et al. 2020). Right panels: Same as left panel, but versus $\log \epsilon (\mathrm{Mg} + \mathrm{Sc} + \mathrm{Ti})/3$. Symbols are chosen as in Fig. 5.9. Different line styles and colors indicate different fitting techniques. LS denotes a least square fitting without any weights. The slope given in each panel is calculated with this fitting technique. A weighted least square fit is indicated with WLS, an orthogonal distance regression is denoted with ODR and a bootstrapping aggregation is denoted with Bagging. Light grey lines indicate different slopes that are involved in the bootstrapping algorithm. The line in the lower right panel indicates a 1:1 correlation without performing a fit.

A linear least squares fit of the lighter heavy element yttrium ($Z = 39$) over magnesium (and a convolved α-abundance consisting of the mean of magnesium, scandium, and titanium abundances.) reveals a slope close to 1, indicating a similar production site of these elements (Fig. 5.20) in all cases with sufficient amount of data (i.e., for Fornax, Sagittarius, and Sculptor). Plotting together Carina, Ursa Minor, and Draco shows a good agreement with a 1:1 correlation (lower right panel of Fig. 5.20). This is a strong indication that yttrium may be produced by fast rotating massive stars (e.g., Pignatari et al. 2008, Frischknecht et al. 2012) or CC-SNe (e.g., Arcones and Montes 2011, Hansen et al. 2014b, Arcones and Bliss 2014). At high metallicities, the s-process will also contribute to the production of yttrium, best visible in the case of Fornax and Sculptor (and only weak indications in Sagittarius, see Fig. 5.20), where the increased scatter at $\log \epsilon (\mathrm{Mg}) > 6$ indicates a contribution from a different process.

The details of the fitted slope depend to some extent also on the fitting procedure. We have applied four different fitting algorithms, one ordinary least squares fit, a weighted least squares fit with each point weighted by $w = 1/\sqrt{x_{\mathrm{err}}^2 + y_{\mathrm{err}}^2}$ (using the python package statsmodel.api.WLS), an orthogonal distance

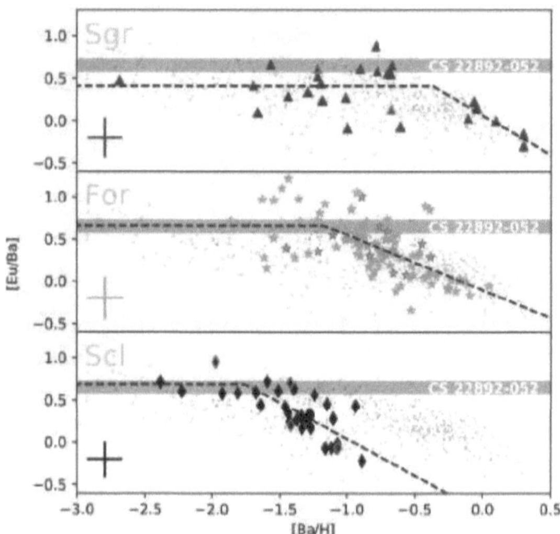

Figure 5.21.: [Eu/Ba] as an indicator of the s-process contribution to Ba. Red symbols indicate stars with unreasonable parallax. The gray band shows the highly r-process enriched star CS 22892-052 (Sneden et al. 2003). Gray dots indicate stars of the MW (Reddy et al. 2003, Cayrel et al. 2004, Reddy et al. 2006). The median error is indicated in the lower left of each panel. This plot was taken from Reichert et al. (2020).

regression including errors in x- and y-direction (using the python package scipy.odr), and a bootstrapping aggregation in combination with a weighted least squares fit. For this, we used the same weights w, taking 80% of all stars per estimator into account (using the python packages sklearn.ensemble.BaggingRegressor and sklearn.linear_model.LinearRegression). The obtained slopes vary only slightly for Fornax, Sagittarius, and Sculptor. Therefore, we only give the value of the slope of a simple unweighted least squares fit (Fig. 5.20). We note, however, that an orthogonal distance regression lead to a constant offset of the slope of ~ 0.2 towards higher values.

In Fig. 5.21, we identify the point where barium is no longer predominantly produced by the r-process but rather by the s-process by using the trend of [Eu/Ba] versus [Ba/H] (cf., e.g., Fig. 20 in Tolstoy et al. 2003 or Fig. 14 in Tolstoy et al. 2009). This is similar to, but less clearly seen in [Eu/Y] versus [Y/H] (see, e.g., Lanfranchi et al. 2008, for a chemical evolution model of this trend). Barium is created in large amounts in the s-process.

Hence, we expect the onset of the s-process to be at metallicities that are similar to those of the α-knee because the s-process is hosted by AGB stars, which are low-mass, slowly-evolving stars in a late phase of their evolution and thus may experience a time delay similar to type Ia SNe (as seen in Sculptor, Hill et al. 2019). The location on the [Ba/H]-axis, where [Eu/Ba] decreases translates to a metallicity (by fitting a linear function with an ordinary least squares fit to [Ba/H] over [Fe/H] of -0.57 dex, -1.04 dex, and -1.57 dex for Sagittarius (cf. Fig. 10 of Hansen et al. 2018), Fornax (cf. Fig. 20 of Letarte et al. 2018 and Fig. 19 of Lemasle et al. 2014), and Sculptor (cf. Fig. 12 of Hill et al. 2019), respectively. Thus, the onset of the s-process seems to occur at metallicities increasing with the stellar mass of the galaxy. For Sculptor it coincides with the position of the α-knee, while for Sagittarius and Fornax this happens at a higher metallicity. This may be affected by uncertainties in determining the locus of the knee. A larger sample size would constrain the onset with more precision, especially in the case of Sagittarius. For low values of [Ba/H], the [Eu/Ba] ratio comes close to a constant value of the r-II[9] star CS 22892-052 (Sneden et al. 2003). In most dSphs, this level is reached at low metallicity. We therefore assume that barium is a clean trace of the r-process for $[\text{Fe/H}] < -2$ and thus $\text{Ba}_r = \text{Ba}$. However, there may be individual stars at low metallicities that still have an s-process contribution from other sites such as rapidly rotating spin stars (e.g., Pignatari et al. 2008, Frischknecht et al. 2012, Chiappini 2013).

5.5.4. The r-process

Recent Galactic chemical evolution (GCE) calculations indicate that the trend of r-process elements with respect to α-elements or iron is thought to be flatter for production sites that have a similar delay time as CC-SNe (for GCE models of the MW see, e.g., Matteucci et al. 2014, Cescutti et al. 2006, van de Voort et al. 2019). GCE models that implement NSMs as the only source of the r-process produce positive slopes $[\text{r}/\alpha]$[10] versus $[\alpha/\text{H}]$ as the production of r-process elements sets in later than the production of α-elements. This can be explained by the larger delay time of NSMs in comparison with CC-SNe. The expected slope is model-dependent and can be influenced by many parameters such as, e.g., the delay time distribution and the rate of the event (cf., Fig. 7 of van de Voort et al. 2019). In addition, the r-process may be hosted by several production sites, which would influence the slope as well. A final answer to distinguish different astrophysical production sites is therefore difficult to obtain and beyond the scope of this work. We can, however, estimate the range of slopes from different galaxies. For this, we look again for correlations of α-elements (represented by magnesium) with $\log \epsilon(\text{Eu})$.

Due to the vanishing europium lines at lower metallicities, the Pearson correlation coefficient (Pearson 1895) is low for all galaxies with more than five stars ($r_{\text{Pearson}} < 0.37$), except for Sagittarius (Fig. 5.23). Supported by the large amount of high-resolution UVES spectra in Sagittarius, we were able to measure europium at lower metallicities. The high correlation coefficient in Sagittarius together with the good metallicity coverage of our sample this results in a highly correlated value of $r_{\text{Pearson}} = 0.92$. Sagittarius is therefore the only galaxy where a linear fit will provide meaningful results. A linear least squares fit of $\log \epsilon(\text{Eu})$ over $\log \epsilon(\text{Mg})$ has a slope of 1.28 ± 0.13, close to a one-to-one correlation, indicating a main production of europium and α-elements on similar timescales at least for Sagittarius (see Fig. 5.23). Rare CC-SNe such as MR-SNe (Winteler et al. 2012, Nishimura et al. 2015, 2017, Mösta et al. 2018) or collapsars (Siegel et al. 2019) may be possible candidates (for a discussion see also Skúladóttir et al. 2019, cf. Fig. 11 of van de Voort et al. 2019). However, as the slope is not 1 within the uncertainty, but more positive, we cannot exclude a delayed contribution to Eu from NSMs.

---

[9]using the definition of Beers and Christlieb (2005)
[10]with "r" being a short writing for an element that traces the r-process.

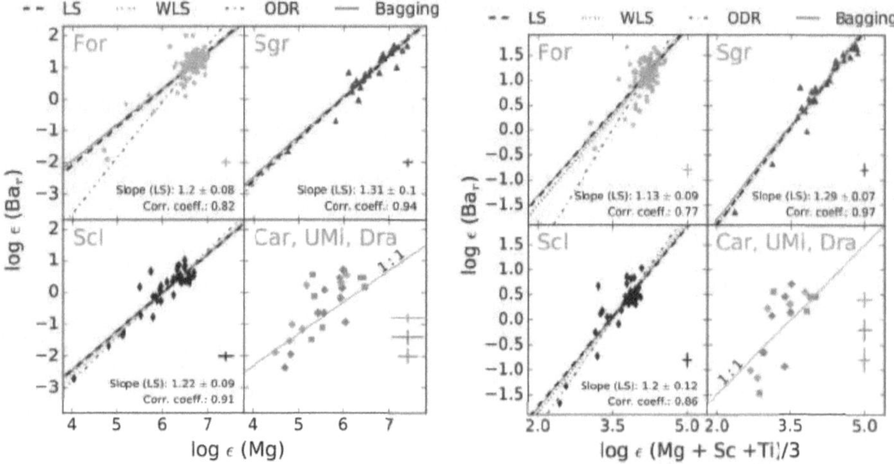

Figure 5.22.: Left panels: Absolute abundance of Ba_r versus absolute abundance of magnesium together with a linear fit (Reichert et al. 2020). Right panels: Same as left panel, but versus $\log \epsilon(Mg + Sc + Ti)/3$. Note that the subscript "r" denotes that Eu was used for high metallicites and Ba for low metallicities trying to filter out the s-process contribution as described in the text. Symbols are chosen as in Fig. 5.9. Different line styles and colors indicate different fitting techniques. LS denotes a least square fitting without any weights. The slope given in each panel is calculated with this fitting technique. A weighted least square fit is indicated with WLS, an orthogonal distance regression is denoted with ODR and a bootstrapping aggregation is denoted with Bagging. Light grey lines indicate different slopes that are involved in the bootstrapping algorithm. The line in the lower right panel indicates a 1:1 correlation without performing a fit.

For barium, we have a larger sample size due to the stronger absorption lines. As previously discussed, we can use barium as a tracer for the r-process below $[Fe/H] < -2$, while at higher metallicities we use europium abundances. To bring both elements to the same scale, we assume that the r-process is robust and europium is only produced by the r-process. We can then convert our europium measurements at $[Fe/H] > -2$ to the r-process fraction of barium abundances by assuming solar r-process residual ratios between these elements (Sneden et al. 2008). We calculate Ba_r as:

$$\log \epsilon(Ba_r) = 1.02 + \log \epsilon(Eu). \tag{5.4}$$

We define

$$[Ba_r/Fe] = (\log \epsilon(Ba_r) - \log \epsilon(Fe)) - (\log \epsilon(Ba)_\odot - \log \epsilon(Fe)_\odot). \tag{5.5}$$

A linear least squares fit of $\log \epsilon(Ba_r)$ and $\log \epsilon(Mg)$ in Fig. 5.22 reveals a slope close to 1, but with a much better correlation coefficient than in the case of $\log \epsilon(Eu)$. This holds for all dSph galaxies where we had sufficient data and therefore in a stellar mass range of $2.1 \cdot 10^7\, M_\odot - 2.9 \cdot 10^5\, M_\odot$ for Sagittarius and Ursa Minor, respectively. Note that Fig. 5.22 does not show Sextans. However, this galaxy is with a slope of 1.39 ± 0.31 also in agreement with the other galaxies. Similar results have been obtained using

$\log \epsilon(\text{Mg}_{\text{NLTE}})$ instead of $\log \epsilon(\text{Mg})$, which demonstrates that the employed NLTE correction does not have a major impact on the derived slopes. Literature values of Reticulum II from Ji et al. (2016b) lead to a slope of 1.11 ± 0.37 indicating a production of α-elements and Ba_r on comparable delay times also in UFD.

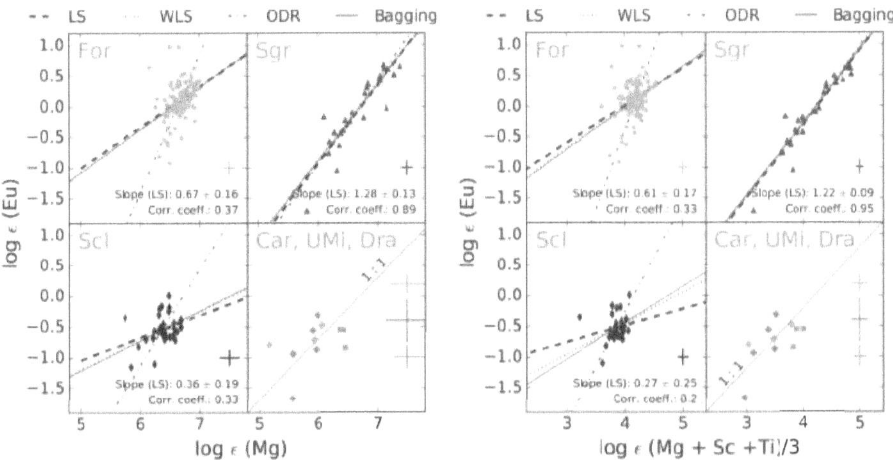

Figure 5.23.: Same as Fig. 5.20, but for Eu.

The occurrence of n-capture elements at very low metallicities causes a problem when arguing for NSMs as the only source of r-process elements. Previous studies (e.g., Argast et al. 2004, Matteucci et al. 2014, Cescutti et al. 2015, Wehmeyer et al. 2015, Haynes and Kobayashi 2019, Wehmeyer et al. 2019) highlighted the problem of producing an early enrichment of n-capture elements at low metallicities $[\text{Fe/H}] < -3$ due to the delay of NSM events as well as the inferred iron abundance that will be mixed in by the previous CC-SNe (since neutron stars are the final product of CC-SNe). Possible solutions to this problem have been suggested such as the pollution of the interstellar medium (ISM) with material from proto-galaxies (e.g., Komiya and Shigeyama 2016). Furthermore, neutron star kicks may move the NSM event far away from the birthplace of the neutron stars and therefore escape the previously synthesized iron (Wehmeyer et al. 2019). Skúladóttir et al. (2019) argued that this would, however, indicate that the MW is as efficient as lower mass dwarfs to maintain the neutron star in the system, which may be unlikely. Another solution could be the occurrence of a black hole-neutron star merger (Wehmeyer et al. 2019).

We confirm the enhancement of [Eu/Mg] for Fornax and Sagittarius at high metallicities as discussed in Skúladóttir et al. (2019) and Skúladóttir and Salvadori (2020). These supersolar values are still an outstanding problem. We note, however, that the metal-rich stars in Sagittarius and Fornax are extremely similar in their stellar parameters. Therefore this issue may be related to a systematic uncertainty when analyzing these stars (e.g., LTE or 3D effects when modelling the structure of the photosphere).

The MW shows a decreasing trend of [Eu/Fe] for high metallicities. Côté et al. (2019a) and Simonetti et al. (2019) argued that this decrease is directly related to a different delay time distribution of NSMs or that there must be an additional source of r-process elements that dominantly contributed in the early universe causing the plateau followed by the declining trend (see chapter 6 for more details). GCE models that assume a similar delay time distribution (e.g., t^{-1}) for both events (i.e., NSMs and type Ia SNe) are unable to reproduce this decreasing trend. All dSph galaxies with sufficient data, namely Sagittarius, Fornax, Sculptor, and Sextans (Fig. 5.12) show this decreasing trend as well. This decrease occurs at similar metallicities as the α-knee (see Table 5.7) and is therefore connected to the additional iron contribution of type Ia SNe. The presence of this decrease in the dSph galaxies favors a similar production site of n-capture elements as in the MW.

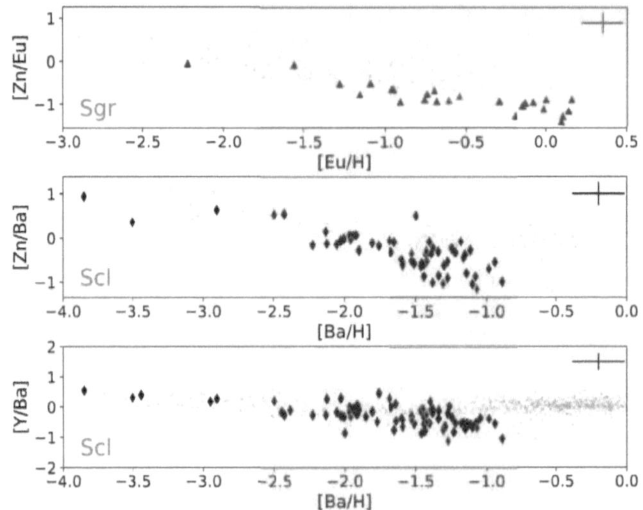

Figure 5.24.: [Zn/Eu] ratio over [Eu/H] for Sagittarius (upper panel), [Zn/Ba] over [Ba/H] for Sculptor (middle panel), and [Y/Ba] over [Ba/H] for Sculptor (lower panel). Gray dots indicate stars of the MW (Reddy et al. 2003, Cayrel et al. 2004, Ishigaki et al. 2013, Fulbright 2000, Nissen et al. 1997, Prochaska et al. 2000, Stephens and Boesgaard 2002, Ivans et al. 2003, McWilliam et al. 1995, Ryan et al. 1996, Gratton and Sneden 1988, Edvardsson et al. 1993, Johnson 2002, Burris et al. 2000). Symbols are chosen as in Fig. 5.9. Stars marked in red indicate stars with close distances according to the parallax. Median errors are indicated at the upper right corner of each panel. (Reichert et al. 2020).

In our study, we determined the abundance of the heavy iron-peak element zinc. The ratios [Zn/Ba] and [Zn/Eu] indicate a plateau at low metallicities, followed by a decreasing trend (Fig. 5.24). As for the α-elements, the ratio is indistinguishable from the MW for low metallicities and drops to sub-solar values for higher metallicities. Though being limited to only two stars in Sagittarius and four stars in Sculptor, the existence of this plateau may suggest a production of zinc, europium, and barium on the same timescales (for values below [Ba/H]~ -2.3). A decrease is also visible in the ratio [Y/Ba]. The locus of the decrease does not agree with the value where [Eu/Ba] decreases (at [Ba/H] ~ -1.6, see Fig. 5.21), so it is most likely disconnected from an s-process contribution. We note, however, that our data may not be sensitive enough to separate between different timescales at this small time interval.

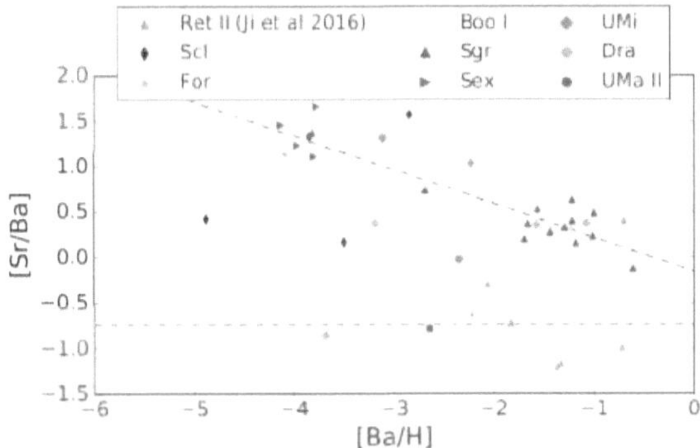

Figure 5.25.: [Sr/Ba] versus [Ba/H]. The green triangle indicates a star of the globular cluster Terzan 7. The dashed red line corresponds to a median of all UFD galaxies, the dashed black line is a linear least square fit to Sagittarius. This plot was taken from Reichert et al. (2020).

We find detectable barium and strontium abundances down to metallicities of [Fe/H] ~ -4 in Sculptor (cf., Kirby et al. 2009, Frebel et al. 2010a), and [Fe/H] $\lesssim -3$ for nearly all investigated dSph galaxies independent of their mass (Fig. 5.12).

The formation of strontium and barium seems to be fundamentally different when comparing UFD galaxies with dSph galaxies or the MW. Mashonkina et al. (2017b) found a deficiency of [Sr/Mg] for UFD galaxies compared to classical dSph galaxies and the MW, whereas [Ba/Mg] follows the same trend in dwarf galaxies of different sizes. They concluded that there are at least two production sites of strontium, one that is coupled to the production of barium and one that is independent from barium. These production sites are present in the MW halo and the dSph galaxies, but the barium-independent production channel of strontium seems to be missing in the UFD galaxies leaving the strontium lower here. Unfortunately, our statistics of strontium in UFD galaxies is limited due to the wavelength coverage of our sample (see Fig. 5.6). However, adding the literature values from Ji et al. (2016b) supports this suggestion.

The UFD galaxies seem to follow a flat trend in [Sr/Ba] over [Ba/H] indicated by a red dashed line in Fig. 5.25 (compare also with Fig. 5 of Ji et al. 2016b and Fig. 8 of Mashonkina et al. 2017b or for the MW with Fig. 13 of Honda et al. 2004). Compared to Mashonkina et al. (2017b), we derive a slightly lower value of [Sr/Ba] $= -0.74$ dex for pure r-process stars. This may be explained by the LTE assumptions in our analysis. Our data suggest that more massive galaxies follow a decreasing trend of [Sr/Ba] with increasing [Ba/H], which points to more than one production channel of strontium. Due to the limited amount of data for UFDs, we cannot draw any firm conclusion about the flat trend of [Sr/Ba] in those galaxies. For this, a larger sample size with strontium measurements is desirable.

6. Evidence for different r-process production sites

In the last chapter, we presented abundances of dSph galaxies that raise hints on different astrophysical hosts for the r-process or a not yet understood early appearance of another r-process site. In the following, we want to focus on the discrepancies that arise when considering NSMs as the only source for the r-process. We do this, by embedding the work that was presented in Côté et al. (2019a) into the knowledge of this book. Therefore, we summarize and condense the most important arguments of this work, but refer the reader to Côté et al. (2019a) for further details. The result of this chapter was a collaborative work together with B. Côté, M. Eichler, A. Arcones, C. J. Hansen, P. Simonetti, A. Frebel, C. L. Fryer, M. Pignatari, K. Belczynski, and F. Matteucci.

6.1. A decreasing trend of [Eu/Fe] in the Milky Way

Similar to the trend in Fig. 5.12, the evolution of [Eu/Fe] in the Milky Way shows a decreasing trend versus metallicity (cyan dots in Fig. 6.1).

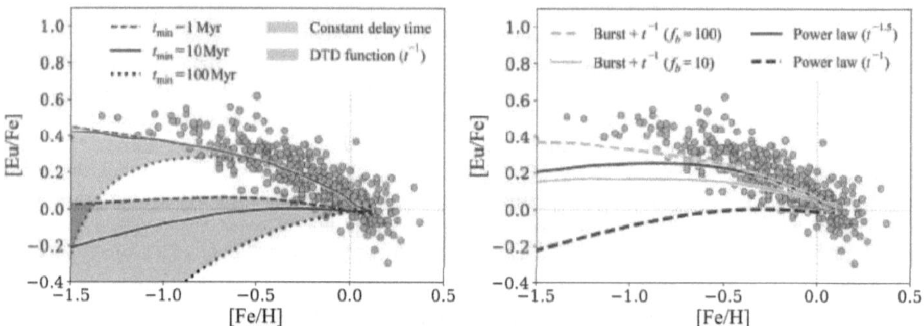

Figure 6.1.: Impact of using different minimum delay times and delay-time distributions for NSMs on the predicted chemical evolution of Eu. Cyan dots are stellar abundances data for disk stars taken from Battistini and Bensby (2016). The thin dotted black lines mark the solar values (Asplund et al. 2009). These figures are taken from Côté et al. (2019a).

From the perspective of Galactic chemical evolution, explaining this trend is a challenge and provides a constrain on the modeling. To be more precise, GCE models that successfully reproduce this trend with NSMs as only source of r-process elements involve a constant delay time instead of a delay-time distribution (DTD). The DTD is defined as the probability of an event to occur at time t after the formation of its progenitor (Fig. 6.2).

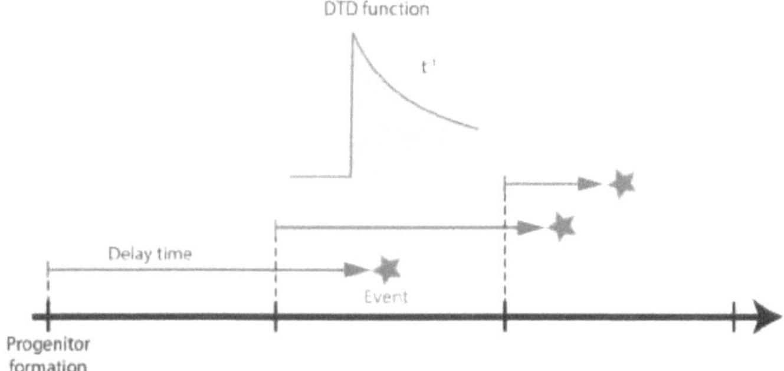

Figure 6.2.: Schematic explanation of the delay time. The delay time is determined by a delay time distribution function, which gives the probability of an event to occur at a given time. Inspired by Fig. 2 of Côté et al. (2019b).

In fact, the problem of modelling the decreasing trend is independent of the complexity of the GCE simulation and is determined by the assumed DTD (see Sect. 3.2 in Côté et al. 2019a). We will discuss the implications of different DTDs together with their observational evidence as well as the impact on the models in the following. Finally, we will condense all constraints from observations on the dominant production channel of r-process elements and summarize problems when assuming different DTDs or hosts for the r-process.

6.2. The delay-time distributions of type Ia SNe and short GRBs

Since the [Eu/Fe] ratio includes iron, also the DTD of type Ia SNe is important to model this ratio. Type Ia SNe are thought to have a DTD in the form of t^{-1} (Totani et al. 2008, Maoz et al. 2010, Graur et al. 2011, Maoz et al. 2012, Perrett et al. 2012, Graur et al. 2014), regardless of the underlying method to derive it (see e.g., Maoz et al. 2014, for an overview of the different methodologies). This form of the DTD for type Ia SNe is also consistent with predictions of population synthesis (Ruiter et al. 2009). We note that there are also suggestions for alternative forms of the DTD as $t^{-1.5}$ (Heringer et al. 2017).

The DTD of NSMs has not been observed so far, but short gamma-ray bursts (GRBs) may be associated with NSMs (e.g., Granot et al. 2017, Lyman et al. 2018, Wu and MacFadyen 2019). This was also inferred by the gravitational wave event GW170817 and the connected detection of GRB 170817A (LIGO Scientific Collaboration and Virgo Collaboration 2017). The DTD for short GRBs may also follow the form of t^{-1} (e.g., Fong et al. 2017), which is also in agreement with population synthesis studies (e.g., Dominik et al. 2012, Chruslinska et al. 2018). However, due to a low number statistics of short GRBs, the form of the DTD function could be different (Fong et al. 2017). As for type Ia SNe, there are suggestions for a steeper DTD in the form of $t^{-1.5}$ (D'Avanzo 2015). However, a steeper DTD of short GRBs should also indicate a steeper DTD of type Ia SNe, indicated by the similar fractions of these events in different types of galaxies (Mannucci et al. 2005, Li et al. 2011, Berger 2014). Furthermore, for 7 of the 13 observations of NS-NS binary systems in the Milky Way there exist estimates of their coalescence time (i.e., 46, 86, 217, 301, 480, 1660, and 2730 Myr, Tauris et al. 2017). The estimated coalescence time of $> 1\,$Gyr for two of them hints towards a long-lasting DTD function.

6.3. Chemical evolution simulations

In the following, we investigate three common approaches to implement NSMs in GCE simulations. First, we assume a constant delay-time as done in e.g., Argast et al. (2004), Matteucci et al. (2014). Afterwards, we assume a DTD function in the form of t^{-1} (e.g., Shen et al. 2015, van de Voort et al. 2015). Finally, we explore alternative forms of the DTD functions. Therefore, we use a simplified version of the GCE code OMEGA (Côté et al. 2017) together with the simple stellar population code SYGMA (Ritter et al. 2018). For all calculations, we assumed a DTD in the form of t^{-1} for type Ia SNe. Additionally, we assumed that NSMs are the only source of r-process elements. By this consideration, the observed robust r-process pattern (see Sect. 2.4.1) can be guaranteed by fission cycling (Beun et al. 2006, Goriely et al. 2011, Roberts et al. 2011, Korobkin et al. 2012, Rosswog et al. 2013, Eichler et al. 2015, Mendoza-Temis et al. 2015) that is caused by the involved high neutron densities. Taking into account that NSMs are rare events, also a large scatter of r-process elements at low metallicities (e.g., of stars in the Galactic Halo) can be explained. In addition, we will discuss if the assumed DTD is able to explain the observation of short GRBs in early-type galaxies (as e.g., GRB 170817A in a S0 galaxy). These galaxies show typically a lack or low level of star formation (e.g., González Delgado et al. 2016) and are dominated by old stellar populations (e.g., van de Sande et al. 2018).

Constant delay time: The model results for the evolution of [Eu/Fe] assuming different constant delay times (i.e., 1, 10, and 100 Myr) are shown as blue band in the left panel of Fig. 6.1. All results are able to describe the observed decrease in [Eu/Fe]. However, the absence of a DTD stands in contradiction with the long-lasting DTD functions that are predicted by population synthesis models (Sect. 6.2). In addition, it is inconsistent with the observation of NS-NS binaries that have an estimated coalescence time, ranging from 46 to 2730 Myr (Tauris et al. 2017). Furthermore, without a DTD it is not possible to explain the detection of short GRBs in early-type galaxies.

Neutron star mergers with a DTD in the form of t^{-1}**:** The DTD in the form of t^{-1} is unable to reproduce the observed decreasing trend of [Eu/Fe] (gray band in the left panel of Fig. 6.1). This long lasting DTD fulfills the constrain from population synthesis models as well as the observed coalescence times of NS-NS binary systems. As the DTD is the same as for type Ia SNe, it can explain that type Ia SNe and GRBs are detected in similar portions in different types of galaxies (see Côté et al. 2019a, for a more detailed discussion).

Neutron star mergers with exploratory DTDs: Neither a constant delay-time nor a DTD in the form of t^{-1} is able to satisfy most observational constraints. Here, we test alternative DTD functions that are shown in the upper panel of Fig. 6.3 (c.f., Simonetti et al. 2019). A steeper DTD in the form of $t^{-1.5}$ is able to reproduce the decreasing trend (black line in the right panel of Fig. 6.1). Considering the uncertainties that are involved when determining the DTD of short GRBs, this DTD may be in agreement with predictions (e.g., D'Avanzo 2015). Also population synthesis models do predict DTD functions steeper than t^{-1}, although the majority are consistent with a t^{-1} distribution. However, assuming a steeper DTD for short GRBs than for type Ia SNe stands in contradiction with the observed similar portions in different types of galaxies.

Alternatively, we assume a burst of NSMs at early times, before the onset of type Ia SNe (orange lines in the right panel of Fig. 6.1 and Fig. 6.3). This allows NSMs to have a wide range of coalescence timescales, in agreement with short GRBs and NS-binary observations, but also reproducing the negative [Eu/Fe] trend. However, assuming a different DTD for short GRBs than for type Ia SNe stands again in contradiction with their observed similar portions in early types of galaxies. We note, that also the dSph galaxies Sagittarius, Fornax, and Sculptor show the same decreasing behavior as the Galactic disk and the burst scenario should therefore also be present in these systems (see Fig. 5.12).

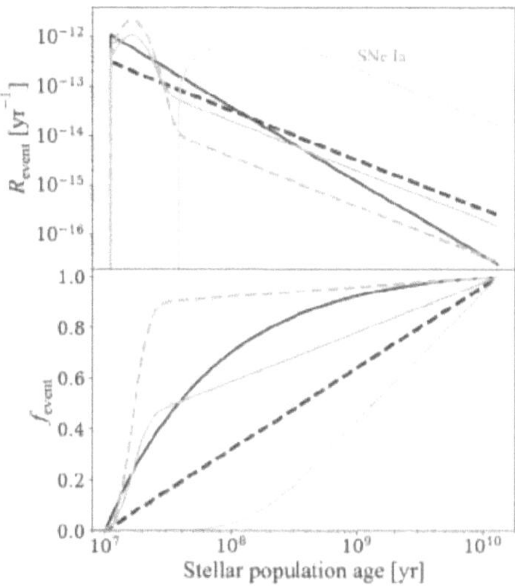

Figure 6.3.: Different delay-time distribution (DTD) functions for NSMs and type Ia SNe. Top panel: Four different DTD functions (black and orange lines) in a simple stellar population of $1\,\mathrm{M}_\odot$. Once integrated, all four DTD functions produce the same number of NS-NS mergers. The grey solid line represents the DTD function used for type Ia SNe. Bottom panel: Cumulated fraction of NS-NS mergers and SNe Ia as a function of time in a stellar population. The dotted horizontal line marks the moment where 50 % of the events have occurred. This figure is taken from Côté et al. (2019a).

6.4. The possibility of an additional source of r-process elements

To summarize, with our current knowledge and assuming NSMs as dominant source of r-process elements it is difficult to explain all presented observational evidences. The above presented inconsistencies are summarized in Table 6.1. There, we used the different DTDs that are described above (i.e., constant delay time, t^{-1}, $t^{-1.5}$, and a burst scenario). For each observational constrain, a *Yes* or a *No* highlights whether or not the selected r-process site is consistent with the constrain. A *Maybe* means that the situation is unclear. The table reflects the current state of understanding. Further observations and improved models could change the interpretation (Côté et al. 2019a).

Table 6.1.: List of agreements and inconsistencies when assuming that current models of NS-NS mergers and MR SNe are the only r-process site for elements in between the second and third peaks. This table is taken from Côté et al. (2019a).

Observational Constraints	NS-NS Mergers			MR SNe
	No DTD	t^{-1} DTD	Modified DTD	
Production of a robust main r-process pattern	Yes	Yes	Yes	Maybe
Possibility of producing actinides elements	Yes	Yes	Yes	Yes
Large scatter of [Eu/Fe] in metal-poor stars	Yes	Yes	Yes	Maybe
Stars with [Eu/Fe] > 0.3 at [Fe/H] $\lesssim -2$	Maybe	Maybe	Maybe	Yes
Decreasing trend of [Eu/Fe] at [Fe/H] > -1	Yes	No	Yes	Yes
Fraction of short GRBs in early-type galaxies	No	Yes	No	–
LIGO/Virgo local NS-NS merger rate density	Yes	Yes	Yes	Maybe
Probability of detecting GW170817 in early-type galaxy	Zero	Low	Very low	–

So far, we did not take into account that there may be two production sites of r-process elements. By assuming a second host event, we are able to reconstruct the decreasing trend of [Eu/Fe] (bottom panel of Fig. 6.4). In Fig. 6.4, the black dashed line indicates the production rate of r-process elements assuming a DTD in the form of t^{-1}. In addition, the figure shows the production rates of iron in CC SNe (blue line) and type Ia SNe (red line) as a function of time in a galaxy with a constant star formation rate of $1 \, M_\odot \, yr^{-1}$. The black lines show the hypothetical production rate of Eu (scaled up by a factor of 4×10^5 for visualization purposes) when assuming that Eu is ejected following the lifetime of massive stars (solid black line) or a delay-time distribution function in the form of t^{-1} (dashed black line), similar to one adopted for SNe Ia. The pink residual dashed line is a possible extra source of Eu in the early universe, obtained by subtracting the black dashed line from the solid black line (Côté et al. 2019a). The purple line is the amount of Eu that we miss with our current understanding and that has to be added to the production rate of NSMs.

Therefore, by assuming that NSMs contribute to the chemical enrichment of europium, this suggests an additional production site of europium that acts at early times in the Galaxy. For consistency, this hypothetical production site should fade away with increasing metallicities and be almost completely negligible today. According to this, NSMs should become the dominant production site, when also type Ia SNe start to dominate over the production of iron ([Fe/H] ~ -1, Battistini and Bensby 2016, Buder et al. 2018).

When integrating the potential contribution of NSMs and the residual contribution, we obtain that approximately 10 % of the current Eu production could come from this hypothetical extra site. This simple approach leads to the conclusion that this production site may has produced 60 % and 40 % of

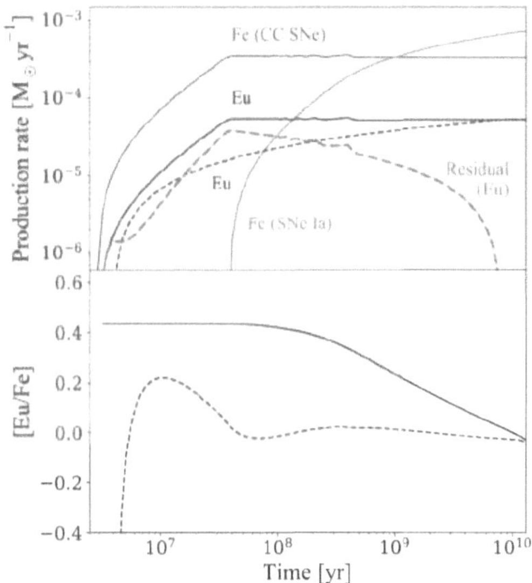

Figure 6.4.: Depiction of how the decreasing trend of [Eu/Fe] can be generated with time. Top panel: Production rate very time for different host events. Bottom panel: Resulting [Eu/Fe] ratio versus time. This figure is taken from Côté et al. (2019a).

the Eu during the first 100 Myr and 1 Gyr of Galactic evolution, respectively. We stress however, that our assumption relies on a DTD function in the form of t^{-1} for NSMs and that our overall production rate is correct. In addition, the DTD of type Ia SNe is an input. Therefore, the here given numbers should not be taken as solid predictions, but rather serve as a motivation for further studies.

We note that also other chemical evolution studies suggest a possible contribution of an additional r-process production site (Argast et al. 2004, Matteucci et al. 2014, Cescutti et al. 2015, Wehmeyer et al. 2015, Haynes and Kobayashi 2019), but with the motivation of injecting r-process elements early enough to explain the Eu abundances in metal-poor stars (see also Safarzadeh et al. 2019). However, different to these studies, our scenario strictly requires that this additional production site fades away as function of the time and metallicity (c.f., purple line Fig. 6.4).

A candidate that would possibly fulfill all constraints are magneto-rotationally (MR) driven supernovae. As summarized in Sect. 2.3.1, these events are thought to play a particular important role in the early universe. In the next chapter, we investigate the nucleosynthesis of such an hypothetical scenario.

7. The production of heavy elements in MR-SNe

All observations hint towards a lack of understanding concerning the production of heavy elements. NSMs seem to either behave differently than commonly thought or there may exist additional astrophysical hosts of the r-process. In the last chapter, we saw that MR-SNe (Sect. 2.3.1) are a promising candidate as they can play a dominant role in the chemical enrichment at early times. In the following, we calculate the nucleosynthesis of this scenario. Modeling the magnetic field is still challenging as pre-collapse models are uncertain and on top of that, the magneto-rotational instability (MRI) adds a particular challenge. Therefore, the magnetic field strength will be a free parameter. This chapter presents the nucleosynthesis analysis of the longest MR-driven SNe calculations with accurate neutrino transport (M1 transport scheme, Just et al. 2015) carried out so far. The underlying hydrodynamical models has been calculated and published in Obergaulinger and Aloy (2020). The presented analysis of the nucleosynthesis is part of an ongoing collaborative work together with M. Obergaulinger, M. Eichler, M. Á. Aloy, and A. Arcones.

7.1. A review of previous studies

Several studies have focused on the nucleosynthesis of MR-SNe. Each of these studies investigated different aspects of the involved uncertainties when modelling MR-SNe (Nishimura et al. 2006, Winteler et al. 2012, Saruwatari et al. 2013, Nishimura et al. 2015, 2017, Mösta et al. 2018, Halevi and Mösta 2018). One of the first studies that post-processed the synthesis of elements in such an event was carried out by Nishimura et al. (2006). Later, Winteler et al. (2012) succeeded with the first nucleosynthesis calculation based on a 3D hydrodynamical simulation (Käppeli 2013). A detailed analysis of five 2D-models, involving different magnetic field strengths, rotation rates, and tracer particle resolutions was carried out by Nishimura et al. (2015). In addition, the impact of neutrinos was investigated by Nishimura et al. (2017). Mösta et al. (2018) suggested that the assumption of 2D-symmetry artificially supports the production of heavy elements and 3D-models need even stronger magnetic fields to successfully synthesize heavy elements. Also a misalignment of the magnetic field with respect to the axis of rotation can have a dominant influence on the nucleosynthesis (Halevi and Mösta 2018).

Our study joins the nucleosynthesis investigations of MR-SNe with a focus on the long-time behavior and sophisticated neutrino transport (c.f. the different studies in Tab. 7.1). Neutrinos play a crucial role in nucleosynthesis calculations as they usually drive matter towards less neutron-rich conditions, because they can change the neutron richness of the irradiated matter. This depends on the energies and luminosities of neutrinos and anti-neutrinos and the equilibrium value can be seen in Appendix B.7. So far, all studies relied on simplified neutrino treatments (e.g., light bulb or leakage schemes, see Tab. 7.1 for an overview). For the first time, we investigate the synthesis of elements by using a spectral two-moment neutrino transport scheme (M1, Just et al. 2015) in the underlying hydrodynamic MR-SNe models (Obergaulinger and Aloy 2020). This causes a broad distribution of neutron-richness, making a whole zoo of nucleosynthetic processes possible. We report the synthesis of heavy elements in neutron-rich matter, but also the production of proton-rich nuclei by the νp-process (Fröhlich et al. 2006, Pruet et al. 2006, Wanajo 2006). In addition, we show that models with a comparably weak magnetic field may also be able to synthesize elements with mass numbers up to A$\sim$ 130. This is possible by following the simulation over a long time until a re-arrangement of the angular momentum changes the shape of the proto-neutron star (PNS), ultimately leading to ejection of neutron-rich matter. This shows the exceptional importance of long-time simulations.

7.2. MHD simulations: code and input physics

We calculate the nucleosynthesis of four of the models that are described by Obergaulinger and Aloy (2020). Even though a detailed description of the models is already given in Obergaulinger and Aloy (2020), we will recapitulate the fundamental details of the models in the following (M. Obergaulinger priv. communication). The simulations of the collapse and the explosion of the stellar cores were performed using the radiation-MHD code ALCAR (Just et al. 2015). The dynamics of the gas and the magnetic field were modeled using the equations of special relativistic magnetohydrodynamics (MHD), closed by the equation of state (EOS) of Lattimer and Swesty (1991) with an incompressibility of $K = 220\,\mathrm{MeV}$ at high densities of $\rho > 6 \cdot 10^7\,\mathrm{g\,cm^{-3}}$ and an EOS based on a gas of electrons, positrons, photons, and baryons (Rampp and Janka 2002). For the baryonic component, the so-called flashing scheme is used which assumes that matter is composed by a mixture of five nuclei, i.e., protons, neutrons, α-particles, Si and Ni. The effects of general relativity are considered by using one of the post-Newtonian TOV potential of Marek et al. (2006).

Neutrinos were treated in the spectral two-moment, or M1, framework derived by expanding the Boltzmann equation of radiative transfer into angular moments of the phase-space distribution function of the neutrinos. This expansion yields balance equations for the energy and momentum densities of the neutrinos. The system is closed by a local algebraic Eddington tensor. We evolve the neutrino moments in the frame co-moving with the gas and include energy-bin coupling terms involving the fluid velocity and gravitational potential (Endeve et al. 2012). Matter and neutrinos couple via the following reactions: emission and absorption of neutrinos by nucleons and nuclei, scattering of nucleons, nuclei, and electrons, electron-positron pair annihilation, and nucleonic bremsstrahlung.

7.3. Supernova models

The simulations are based on model 35OC for a star of an initial mass $M_{\mathrm{ZAMS}} = 35\mathrm{M_\odot}$ from the stellar-evolution calculations by Woosley and Heger (2006). Rotation and magnetic fields were included in the spherically symmetric models following the recipe of Spruit (2002). Within the series of four models to which this progenitor belongs to, the mass loss was a free parameter. We selected the model with the second smallest mass loss, which at collapse has a mass of $M = 28.1\mathrm{M_\odot}$ and possesses an iron core of mass $M_{\mathrm{Fe}} = 2.02\mathrm{M_\odot}$ as our reference model (35OC-RO). This choice is motivated by the presence of higher rotation rates with lower mass loss. Previous studies have chosen models with larger mass loss (Tab. 7.1), so called Wolf-Rayet stars that are thought to possibly be the progenitors of hyperenergetic type Ib/Ic SNe. The choice of a Wolf-Rayet star as progenitor implies almost negligible rotation rates (e.g., model E25 of Heger et al. 2000, used in Mösta et al. 2018) that are sometimes artificially set to

Table 7.1.: Summary of basic properties of previously performed MHD simulations.

Study	Model	M_{ZAMS} [$M_\odot$]	M_{bc} [$M_\odot$]	Dimension	b [G]	t_f (pb) [ms]	# Tracers (ejected/total)	Neutrinos	Max. r-process peak
Nishimura et al. (2006)	1	13	3.3	2D	0	310	-/5000	None	2nd
	2	13	3.3	2D	$5.4 \cdot 10^{12}$	433	-/5000	None	3rd
	3	13	3.3	2D	$1.0 \cdot 10^{13}$	354	0/5000	None	No expl.
	4	13	3.3	2D	$5.2 \cdot 10^{13}$	366	-/5000	None	3rd
Winteler et al. (2012)	1	15	11.9	3D	$5.0 \cdot 10^{12}$	33	136/20005	Leakage	3rd
Saruwatari et al. (2013)	1	13	3.3	2D	0	172	-/24000	Leakage	-
	2	13	3.3	2D	$5.4 \cdot 10^{12}$	186	-/24000	Leakage	-
	3	13	3.3	2D	$1.0 \cdot 10^{13}$	200	-/24000	Leakage	-
	4	13	3.3	2D	$5.2 \cdot 10^{13}$	300	-/24000	Leakage	1st
	5	13	3.3	2D	$5.2 \cdot 10^{13}$	444	-/24000	Leakage	3rd
Nishimura et al. (2015)	1	25	5.45	2D	$1.0 \cdot 10^{11}$	103	-/50000	Leakage	2nd
	2	25	5.45	2D	$1.0 \cdot 10^{11}$	58	-/50000	Leakage	3rd
	3	25	5.45	2D	$1.0 \cdot 10^{12}$	57	-/50000	Leakage	3rd
	4	25	5.45	2D	$1.0 \cdot 10^{12}$	77	-/50000	Leakage	3rd
	5	25	5.45	2D	$1.0 \cdot 10^{12}$	33	-/50000	Leakage	3rd
Nishimura et al. (2017)	1-9	15	11.9	2D	$2.0 \cdot 10^{11}$	$\geq$235	-/-	Light-bulb/parametrized	2nd and 3rd(mod. ν)
Mösta et al. (2018)	1	25	5.45	sym. 3D	$1.0 \cdot 10^{12}$	~100	-/100000	Leakage	2nd(or. ν) and 3rd(mod. ν)
	2	25	5.45	3D	$1.0 \cdot 10^{12}$	~110	-/100000	Leakage	2nd(or. ν) and 3rd(mod. ν)
	3	25	5.45	sym. 3D	$1.0 \cdot 10^{13}$	~20	-/100000	Leakage	3rd (or. and mod. ν)
Halevi and Mösta (2018)	1	25	5.45	sym. 3D	$1.0 \cdot 10^{13}$	~20	-/20000	Leakage	3rd
	2	25	5.45	sym. 3D	$1.0 \cdot 10^{13}$	~22	-/20000	Leakage	3rd
	3	25	5.45	sym. 3D	$1.0 \cdot 10^{13}$	~28	-/20000	Leakage	3rd
	4	25	5.45	sym. 3D	$1.0 \cdot 10^{13}$	~49	-/20000	Leakage	2nd
This chapter	1	35	28.1	2D	Or.($\sim 10^{10} - 10^{11}$)	1600	6570/204800	M1	1st
	2	35	28.1	2D	$1.0 \cdot 10^{10}$	2540	7272/204800	M1	2nd
	3	35	28.1	2D	$1.0 \cdot 10^{12}$	900	17446/204800	M1	3rd
	4	35	28.1	2D	$1.0 \cdot 10^{-6} \cdot$ Or.	1530	2218/204800	M1	Fe

Note: Entries that are indicated with "-" represent missing information. The first column indicates the reference to the corresponding study, the column "Model" assigns a number to the models presented there. The columns "M_{ZAMS}" and "M_{bc}" indicate the mass of the progenitor during the zero age main sequence and before collapse, respectively. The dimension (2D-axisymmetric or 3D) is shown in column five. There, "sym. 3D" indicates 3D simulations without introducing seed perturbations which can be compared to octant symmetric simulations. The inferred initial magnetic field is shown in the column "b [G]". An ordinary field strength "Or." indicates the radial dependent profile of the progenitor. Column "t_f (pb)" shows the final simulation time after bounce. The amount of ejected and total placed tracer particles is indicated in "# Tracers (ejected/total)". The final two columns, "Neutrinos" and "Max. r-process peak" show the treatment of neutrinos in the hydrodynamic-simulation and the obtained maximum synthesized elements, respectively. In the ninth column, "mod. ν" indicates that the neutrino properties are modified in order to obtain the given result. The final column shows the approximate final abundance result.

higher values. Model 35OC-RO rotates differentially with an angular velocity $\Omega_c = 1.981/\mathrm{s}$ at the center and $\Omega_{\mathrm{Fe}} \approx 0.11/\mathrm{s}$ at the surface of the Fe core. The data contain the radial profiles of the poloidal and toroidal components of the magnetic field in the radiative layers (Fig. 7.1). In convectively unstable layers, the field is set to zero by construction. With a field strength of $b^{\mathrm{pol;tor}} \approx 1.7 \cdot 10^{10}; 1.7 \cdot 10^{11}$ G for the poloidal and toroidal components at the center of the core, the model possesses a relatively, though not extremely, high magnetization. We constructed the 2D distribution of the magnetic field from these radial profiles by assuming a sine dependence in θ.

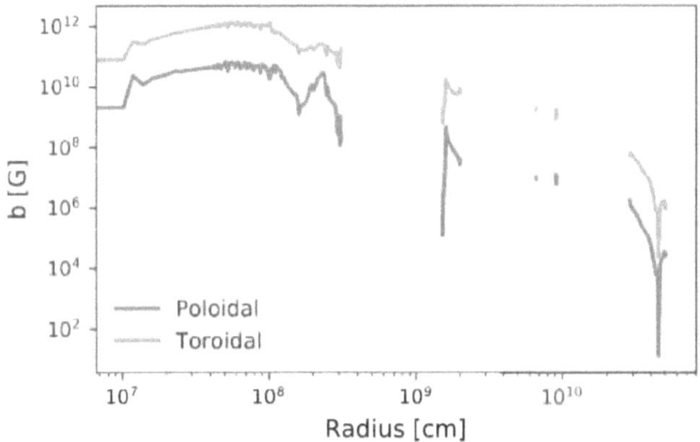

Figure 7.1.: Radial profile of the magnetic field strength of our reference model (35OC-RO). Shown are the poloidal (blue) and the toroidal (orange) component. Note that the field is set to zero in convectively unstable regions of the star.

Obergaulinger and Aloy (2020) varied the original profiles of the rotational velocity and the magnetic field of model 35OC-RO to set up the other three models presented here. Models 35OC-Rw and 35OC-Rs are based on the same rotational profile, but replacing the magnetic field by an artificial distribution of poloidal and toroidal field following the prescription of Suwa et al. (2007). The normalisation of the field strengths are $b^{\mathrm{pol/tor}} = 10^{10/10}$ G for 35OC-Rw and $b^{\mathrm{pol/tor}} = 10^{12/12}$ G for model 35OC-Rs. Model 35OC-RRw has an initial field that is six orders of magnitude weaker than that of 35OC-Rw and thus dynamically insignificant. Model 35OC-RRw rotates 1.5 times as fast as model 35OC-Rw. The four models evolve in fairly different ways (see Table 7.2 for an overview of the models).

Model 35OC-Rw develops a neutrino-driven explosion after about 400 ms post bounce with a dynamically unimportant magnetic field at that time. Driven by strongly anisotropic neutrino fluxes, the explosion has the form of a relatively wide bipolar outflow. The shock wave reaches a polar radius of $R \approx 3 \cdot 10^4$ km at $t \approx 2$ s. At that point, the ejecta contain an energy of $\approx 6 \cdot 10^{50}$ erg and a mass of $\approx 0.2 \mathrm{M}_\odot$. The proto-neutron star (PNS) grows in mass by accretion to a baryonic mass of $M \approx 2.4 \mathrm{M}_\odot$, but is stabilised against gravitational collapse by its high rotational energy of $E^{\mathrm{rot}} \approx 8 \cdot 10^{52}$ erg. The magnetic field in and around the PNS grows continuously, but experiences a particularly strong increase after $t \sim 1.8$ s. This growth causes a more efficient redistribution of angular momentum from the central regions to the outer layers. Consequently, growing centrifugal support leads to an expansion of the PNS at low latitudes.

Table 7.2.: Overview of investigated CC-SNe models.

Name	Rotation	Magnetic field	t_{exp} ms	E_{exp} B	M_{ej} $10^{-1}M_\odot$
35OC-RO	Or	Or	178	0.87	1.40
35OC-Rw	Or	10,10	378	2.80	3.91
35OC-Rs	Or	12,12	20	4.16	3.89
35OC-RRw	$\times 1.5$	'Or' $\cdot 10^{-6}$	343	0.209	0.345

Note: We list the name of the model (first column), the initial conditions for rotation and magnetic field (second and third column), the time at which the explosion starts (fourth column), the final energy and mass of the unbound ejecta (fifth and sixth column), and the number of ejected and total tracer particles (final column). In the second column, 'Or' stands for models in which we use the original rotational profile of the progenitor as given by the stellar evolution modeling and $\times 1.5$ indicates that we multiply this rotational profile by a factor of 1.5. The model in which we also used the original magnetic field is denoted with 'Or' in the column 'Magnetic field'. 'Or $\cdot 10^{-6}$' indicates that we reduced the same field by a factor of 10^{-6}, making it dynamically insignificant. In the same column, a pair of numbers, a, b, stands for models in which we replaced the original field by a dipolar magnetic field with a normalization of 10^a G and 10^b G for the poloidal and toroidal components, respectively.

The increasingly oblate PNS extends into surrounding neutron-rich gas, with important consequences for the conditions of matter ejected.

Model 35OC-RO explodes earlier ($t \approx 200$ ms) due to the strong magnetic forces that play an important role in accelerating the outflows. They furthermore lead to a stronger collimation than in model 35OC-Rw. The explosion is faster and more energetic, reaching the radius of $3 \cdot 10^4$ km and an energy of $6 \cdot 10^{50}$ erg about 700 ms earlier than 35OC-Rw, whereas the evolution of the ejected mass is very similar in both cases. The PNS is even more massive than in the previous model with $M \approx 2.6 M_\odot$ at $t \approx 2$ s. Compared to model 35OC-Rw, the magnetic field in the PNS is stronger to begin with. Hence, efficient angular-momentum redistribution and the associated high axis ratio of the PNS develop earlier.

Model 35OC-Rs explodes almost immediately after core bounce. The explosion, driven entirely by the extremely strong magnetic fields, achieves a maximum radius of $3 \cdot 10^4$ km already at $t \approx 0.7$ s. At that time, the energy and mass of the ejecta are $E \approx 3 \cdot 10^{51}$ erg and $0.35 M_\odot$, i.e., they grow strongest of all models. The PNS on the other hand has the lowest mass, because the strong explosion shuts down accretion. It reaches $M \approx 1.9 M_\odot$ at $t \approx 0.5$ s and afterwards tends to slowly lose mass. Without further accretion, the PNS also does not gain angular momentum, resulting in a comparably low rotational energy $E^{\mathrm{rot}} \approx 2 \cdot 10^{52}$ erg at $t \approx 0.7$ s.

Model 35OC-RRw explodes at about the same time as 35OC-Rw, though less violently. At the end of the simulation ($t \approx 1.5$ s), the maximum shock radius is $R \approx 1.5 \cdot 10^4$ km, and the ejecta energy and mass are $E \approx 2 \cdot 10^{50}$ erg and $0.03 M_\odot$, respectively, i.e., considerably less than 35OC-Rw at the same time. The reason for this weaker explosion is that the high rotational energy ($E^{\mathrm{rot}} \approx 120 \cdot 10^{51}$ erg at $t = 1.5$ s) reduces the accretion luminosity and, consequently, the neutrino heating rate. On the other hand, rotation allows for an exceptionally high PNS mass of $M \approx 2.65 M_\odot$ and a high axis ratio.

7.4. Tracers and nucleosynthesis calculation

The evolution of the ejecta is followed by 204800 Lagrangian tracer particles (Sect. 3.4) that are set up at the beginning of the simulations. Into each grid cell, we insert $n_{nc} = 4$ tracer particles at random positions. Each particle represents a fraction of n_{nc}^{-1} of the total mass of the cell, $m_{cell} = \int_{cell} dV \rho$, where ρ is the local mass density. Consequently, the distribution of particle masses is non-uniform and biased towards regions of high density. This disparity is reduced by the logarithmic spacing of the radial grid as zones at higher radii have in general both larger volumes and lower densities.

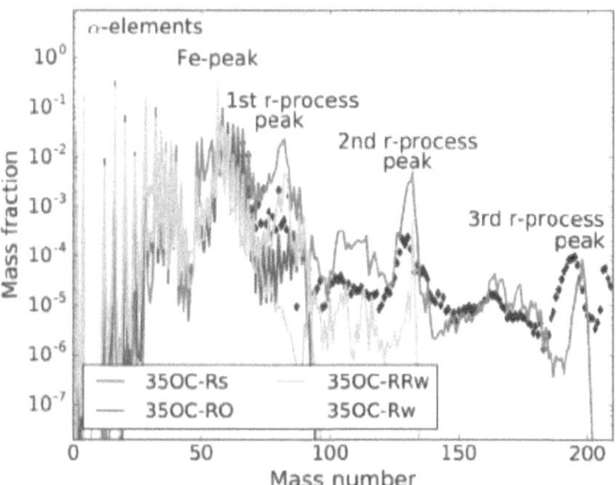

Figure 7.2.: Mass integrated nucleosynthetic yields for the models outlined in Sect. 7.3. For comparison, black diamonds indicate the rescaled solar r-process residual abundances.

We use WinNet (Winteler 2014, Winteler et al. 2012, see also Sect. 3.4) for the nucleosynthesis calculation. The nucleosynthesis calculations start when the temperature of the tracers drops below $T = 20$ GK. We assume nuclear statistical equilibrium (NSE, see Sect. 3.1) for 20 GK$> T >$7 GK and evolve weak reactions and the corresponding Y_e variation[1]. If the maximum temperature of a tracer is below 7 GK, the initial composition is determined by the composition of the progenitor. For $T < 7$ GK, the full network gives the detailed evolution of the abundances of each isotope until 1 Gyr, when most of the nuclei have decayed to stability. We extrapolate from the last time of the tracers assuming an adiabatic expansion and density evolution as $\rho \propto t^{-3}$ (see Appendix A).

---

[1] We observe deviations between the Y_e from the hydrodynamical simulations and the one calculated in the network. This can become significant and depends on the initial temperature. Starting at a high temperature of 20 GK reduces these discrepancies.

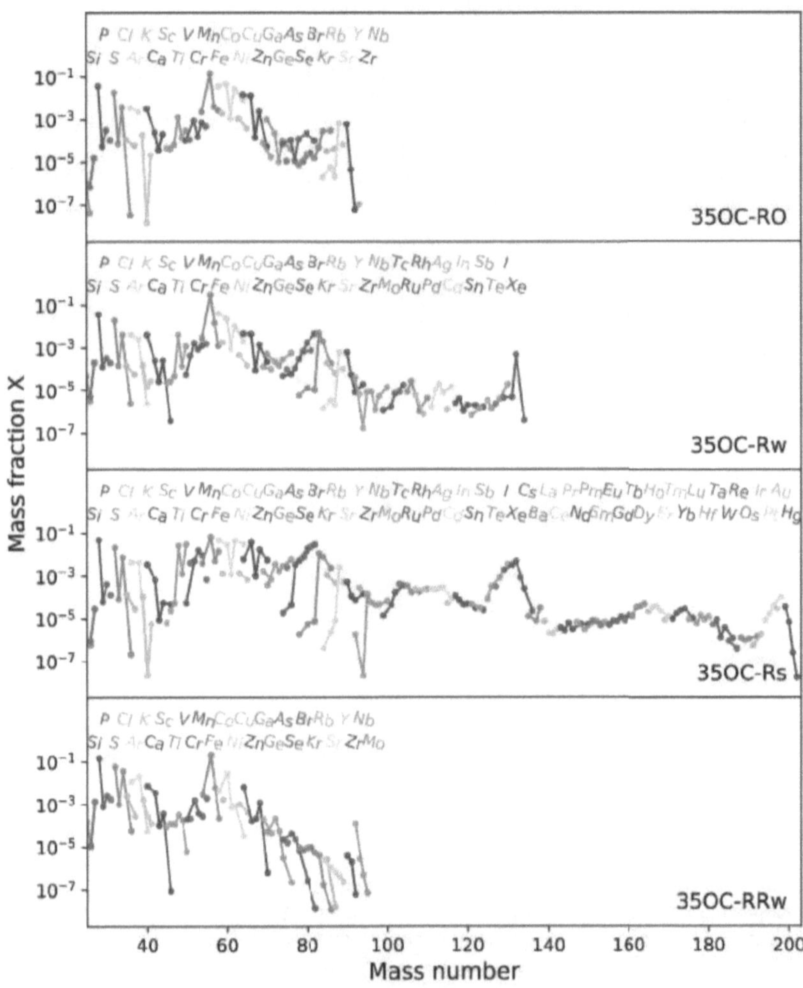

Figure 7.3.: Final mass fractions of individual isotopes for every model. Same elements are indicated by colors and connected by a line. The element names are given at the top of each panel. Nuclei with mass fractions $\leq 10^{-8}$ are not shown.

7.5. Ejecta dynamics and nucleosynthesis

Figure 7.2 presents the mass integrated nucleosynthesis pattern for the models introduced in Sect. 7.3. The mass fraction of individual isotopes is depicted in Fig. 7.3. The differences in the abundances indicate that these models cover a wide range of nucleosynthesis conditions allowing to explore the impact of rotation, magnetic fields, and neutrinos. The models 35OC-RO and 35OC-RRw are similar to typical supernova explosions and produce also "standard" nucleosynthesis, namely elements up to the iron group and lighter heavy elements around $A \sim 90$ (e.g., Harris et al. 2017, Eichler et al. 2018, Wanajo et al. 2018). The model with strong magnetic fields (35OC-Rs) produces elements up to the third r-process peak (see also Nishimura et al. 2006, Winteler et al. 2012, Saruwatari et al. 2013, Nishimura et al. 2015, 2017, Mösta et al. 2018, Halevi and Mösta 2018). The model 35OC-Rw is peculiar due to long-time evolution features that trigger the late ejection of neutron-rich material (Sect. 7.5.2).

7.5.1. Nucleosynthesis patterns and hydrodynamical conditions

In order to understand the integrated abundances, we explore the hydrodynamical conditions of individual tracer particles and their corresponding contribution to the integrated abundances. The composition of every tracer particle is illustrated in the left panels of Fig. 7.4. Groups of tracer particles with similar hydrodynamical conditions (i.e., evolution of temperature and density as well as neutron richness) lead to similar abundance patterns. We have separated these groups with help of a k-mean clustering algorithm (Lloyd 1982)[2] for the abundances. The six groups are indicated by different colors and the average composition of each is shown in the right panels of Fig. 7.4.

The hydrodynamic properties that lead to the observed nucleosynthesis patterns are illustrated in the left and right panel of Fig. 7.5. Every panel of these figures corresponds to one of the four models (see Table 7.2) and every dot corresponds to a tracer with the colors being the same as for the abundances in Fig. 7.4. Whereas the maximum temperature and density separates the groups that contain the lightest elements (right panel of Fig. 7.5), hotter tracer particles are separated by entropy, electron fraction and timescales (Fig. 7.5). Although the classification has been done based on abundance patterns, the groups have a clear dependency on the electron fraction. The histograms indicate the mass distributions of the ejecta in dependence of entropy and electron fraction. All models are dominated by symmetric matter (i.e., $Y_e = 0.5$). To illustrate the clear dependency of the final synthesized elements on the neutron-richness, we took an exemplary trajectory at a given electron fraction and show the final nucleosynthetic pattern in Fig. 7.7. A large amount of neutrons makes the synthesis of heavier elements possible. However, conditions with large numbers of protons (e.g., $Y_e = 0.6$) enable the νp-process (Pruet et al. 2006, Fröhlich et al. 2006, Wanajo 2006) and can therefore also lead to the synthesis of elements that are heavier than iron (c.f. the abundance patterns at $Y_e = 0.5$ and 0.6 in Fig. 7.7).

In addition to the entropy and electron fraction, the expansion time scale also affects the nucleosynthesis. One can get an estimate about the expansion velocity of the different groups in Fig. 7.8. These figures show the radii of tracers in model 35OC-Rs. The colors indicate the nucleosynthesis groups as in Fig. 7.4. The tracers with more neutron-rich conditions reach smaller radii close to or even in the neutron star. Moreover, the expansion can be very fast as shown by the rapid increase of the radius. This fast expansion prevents neutrinos to react with neutrons into protons and preserve neutron-rich conditions of the ejecta.

---

[2]using the python implementation of Pedregosa et al. (2011)

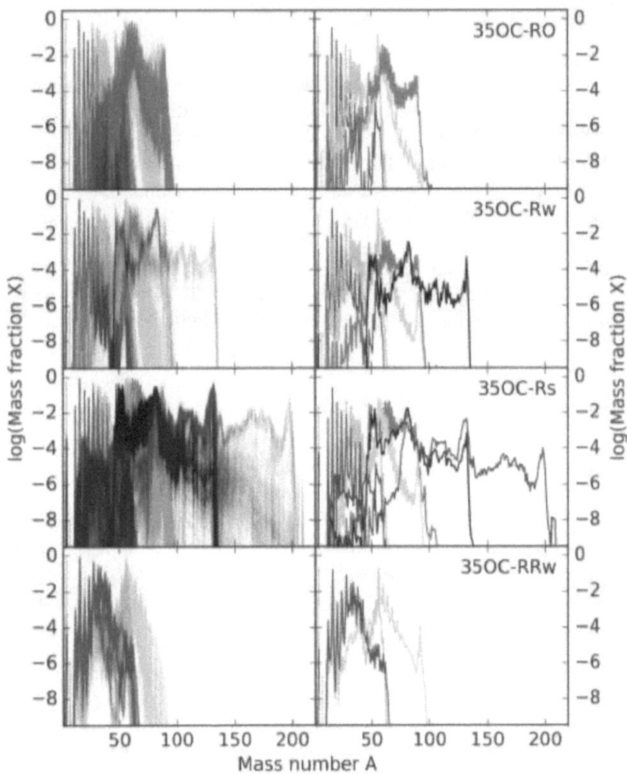

Figure 7.4.: Composition of all calculated tracer particles, indicated by one color for each group, binned by a k-means clustering algorithm. Left panels: Final mass fractions after decay for each individual tracer. Right panels: Mass weighted integrated composition separated into the different groups.

In total, we binned the ejecta into six different groups. The hydrodynamic conditions of them are described in the following:

The α **group** (blue) contains tracers that are located at large radii, which means that they encounter the shock at late times, just before the end of the simulation (Fig. 7.11). Therefore, their peak density and temperature do not exceed $\rho = 10^6 \ \mathrm{g\,cm^{-3}}$ and $T = 3$ GK, respectively, and their final nuclear composition resembles the original progenitor values, i.e. mainly $\alpha-$elements (right panel of Fig. 7.5).

The α-**Fe group** (orange) reaches higher peak densities and temperatures than the α group ($\rho_{\mathrm{peak}} \approx 10^7 \ \mathrm{g\,cm^{-3}}$ and $T_{\mathrm{peak}} \approx 5$ GK), leading to enhanced iron-group abundances due to explosive burning, alongside a considerable amount of $\alpha-$elements.

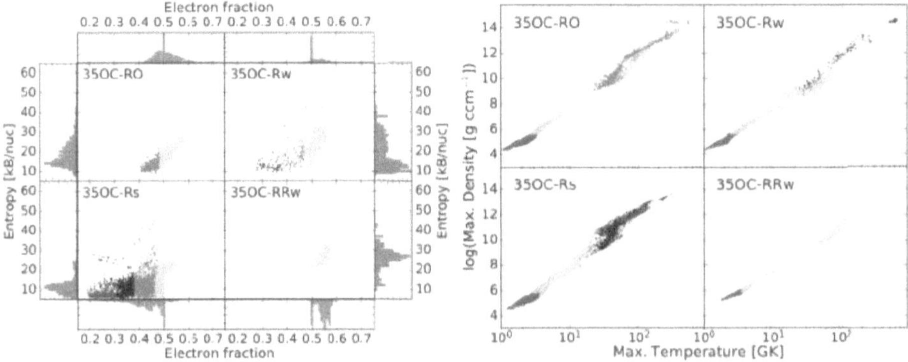

Figure 7.5.: Left Panels: Electron fraction and entropy for each tracer at $T = 5.8$ GK together with their mass weighted distributions at the outermost panels. Colors indicate the corresponding bin of the tracer with the same color code as Fig. 7.4 and Fig. 7.8. Note that we only include tracer particles that reach a peak temperature of at least 5.8 GK here. Right panels: Maximum temperature and density of each tracer particle. The colors indicate nucleosynthetic groups as described in Sect. 7.5.1.

These two groups (α and α-Fe) do not reach NSE conditions and they are characterised by their peak temperature and density, visible by the clear separation of the groups in the right panel of Fig. 7.5. With ongoing simulation time, more progenitor material will be ejected. This is, however, not reflected in our integrated nucleosynthesis, because we are not applying any integration to the outer layers of the star after the end of the simulation.

The **Fe group** (green) comprises two components: (a) mass elements close to the center at relatively large angles from the poles with low expansion velocities and (b) tracers at intermediate radii in the polar directions (see left panel of Fig. 7.5 and Fig. 7.8). Their electron fractions lie between $0.48 \lesssim Y_e \lesssim 0.6$ (left panel of Fig. 7.5), with the highest Y_e values in the center of the jets (see right panel of Fig. 7.5), most pronounced in the reference model 35OC-RO. The angle dependence of the electron fraction is also visible in Fig. 7.6, where we show the simulation time, angle, and electron fraction for every tracer at 5.8 GK. Under these conditions, nuclear reactions favor the production of ^{56}Ni which later decays to ^{56}Fe. In addition, trans-iron nuclei up to Zr and Mo are formed (Fig. 7.3, see also Sect. 7.6). This group is diverse with respect to minimum radius. It contains slightly shocked material, but also fast ejected material that came close to the neutron star (Fig. 7.8). The latter is more exposed to neutrinos and therefore reaches higher electron fractions and produces p-nuclei.

The **Fe-weak-r-process group** (red) differs from the Fe group in the electron fraction, with $0.38 \leq Y_e \leq 0.48$ (left panel of Fig. 7.5). Most tracers of this group start at small radii, experiencing a strong and early shock heating (with the exception of Model 35OC-Rw, see Sect. 7.5.2.). Due to high densities at freeze-out from NSE the composition at this time consists mainly of iron-group nuclei which serve as seed nuclei for the free neutrons that are available at moderate abundances. The resulting final composition is dominated by $50 \leq A \leq 100$ nuclei (see Fig. 7.4).

The **weak r-process group** (black) exhibits similar peak conditions with even lower electron fractions compared to the Fe-weak-r-process group (Fig. 7.5). The lower Y_e values originate from a faster expansion and smaller exposure to neutrinos in the expansion phase. As a result, nuclei up to mass numbers around $A = 140$ (i.e., the second r-process peak) are produced, corresponding to a weak r-process.

Finally, the **r-process group** (purple) contains fluid elements originating close to the surface of the PNS which experience quick acceleration along the jets shortly after bounce. At the end of the simulation, this matter is located in a cocoon around the jet (Fig. 7.11). Experiencing only weak neutrino irradiation, the electron fraction stays low during the expansion ($Y_e \leq 0.3$; see left panel of Fig. 7.5) and a full r-process up to the third r-process peak ($A \approx 195$) is enabled. While all other groups are still synthesized until the end of the simulation, the r-process group is completely determined by the first moments after bounce and the amount of this matter will not change with simulation time (Fig. 7.8 and Fig. 7.6).

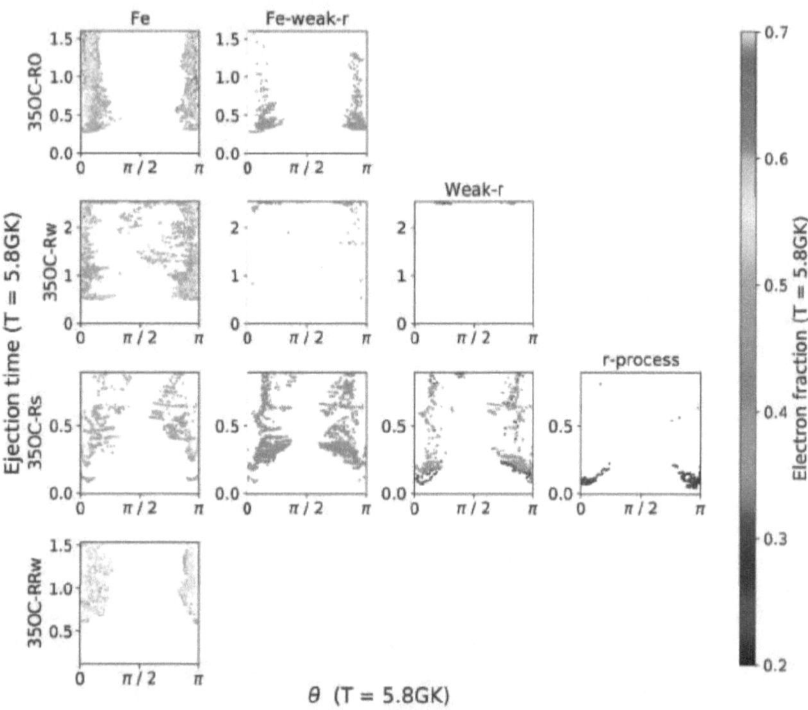

Figure 7.6.: Ejection time (after bounce), angle, and electron fraction for every tracer at $T = 5.8$ GK. The different panels separate the hydrodynamical models (rows) and the different nucleosynthetic groups (columns).

These groups are similar in all models but not every model contains all groups. An exception is the ejection mechanism of model 35OC-Rw that is discussed in Sect. 7.5.2. The α and α-Fe groups are present in all models, because they consist of fluid elements that correspond to slightly shocked progenitor material.

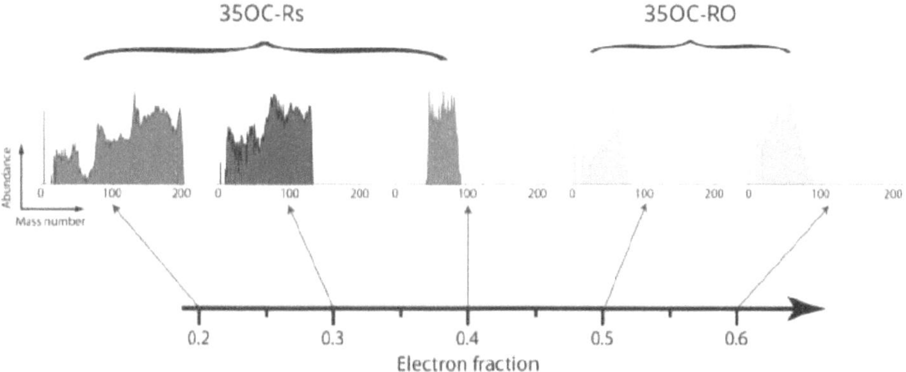

Figure 7.7.: Nucleosynthetic patterns for individual tracers at different electron fractions (at $T = 5.8\,\mathrm{GK}$). The right two patterns are taken from model 35OC-RO, whereas the more neutron-rich patterns are taken from model 35OC-Rs.

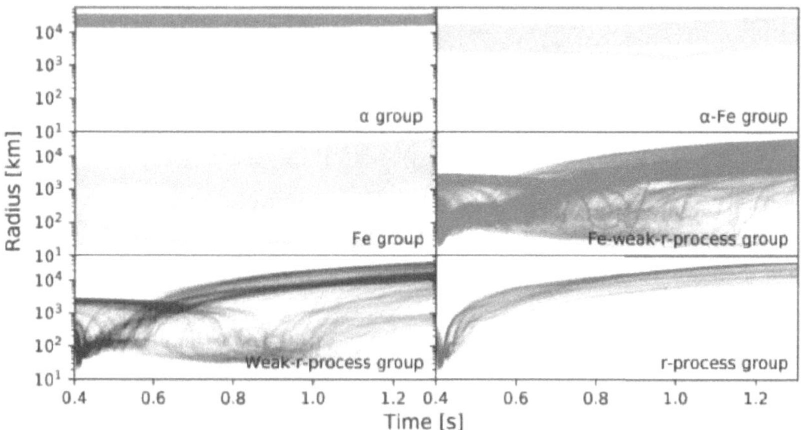

Figure 7.8.: Time evolution of radii for all tracers in model 35OC-Rs. The different panels separate different nucleosynthetic groups that are outlined in Sect. 7.5.1.

7.5.2. A comparison of model 35OC-Rw and 35OC-RRw

The effect of rotation can be investigated by comparing the two models with similar weak magnetic fields: 35OC-Rw and 35OC-RRw. Both models have similar abundances for α-elements up to the iron group. Note, however, that outer shells of the star may still contribute to the α-elements as we did not integrate the progenitor outwards[3].

Model 35OC-RRw with strong rotation and weak magnetic field contains only proton-rich and symmetric matter in addition to the α-groups. Rotation reduces the accretion and thus the accretion luminosity which makes the explosion slower, weaker and keeps matter exposed to neutrinos for longer times. The result is a proton-rich ejecta as shown in the left panel of Fig. 7.5. The majority of tracers never reach the PNS and stay during long time at a similar radius.

The model with slower rotation (35OC-Rw) and stronger explosion produces elements around Sr, Y, Zr ($A \sim 90$, corresponding to the closed neutron shell at $N = 50$) and even low abundances up to the second r-process peak ($A \sim 130$). Most of the matter is ejected with $Y_e \sim 0.5$, a small amount is slightly neutron rich and the weak r-process produces the lighter heavy elements up to around Ag.

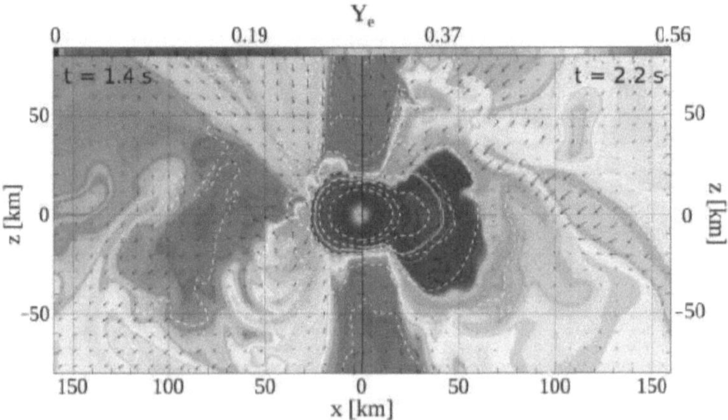

Figure 7.9.: Left panel: Electron fraction of model 35OC-Rw in a region around the PNS at $t \sim 1.4$ s with neutrinospheres in colors and constant densities of 10^{14} to 10^{10} g cm^{-3} in white dashed lines. The arrows indicate the velocity of the fluid. Right panel: Same for $t \sim 2.2$ s. Courtesy of M. Obergaulinger.

Moreover, a small fraction of matter is ejected late and with a large amount of neutrons, $Y_e \sim 0.3$. This is clearly visible by the radial evolution of the group "weak r-process" in Fig. 7.10. This group is composed of tracers ejected at very late times $t \gtrsim 2$ s from the immediate vicinity of the PNS. The sudden appearance of such a population of tracers is the consequence of a relatively abrupt change in the PNS structure that occurs slightly earlier. Up to $t \sim 1.4$ s, the PNS is almost spherical with a decreasing radius and an aspect

---

[3]typically by a mass cut or an angle dependence, see, e.g., Eichler et al. 2018

ratio close to unity (left panel of Fig. 7.9, where the violet contours representing the neutrinospheres can be taken as proxies for the PNS surface). Eventually, however, its magnetic field grows sufficiently to redistribute angular momentum to the outer layers. The excess centrifugal support causes these layers to expand and leads to a growth of the ratio between equatorial and polar radius beyond a value of two (see the right panel of Fig. 7.9). This expansion affects matter of very low Y_e (in Fig. 7.9, marked by blue colors), some of which even ends up outside of the neutrinospheres. The turbulent fluid flows in this region stochasticly advect parcels of this very neutron-rich matter into the polar outflows. These fluid elements will be ejected with high velocities and Y_e therefore stays low (left panel of Fig. 7.5). We note

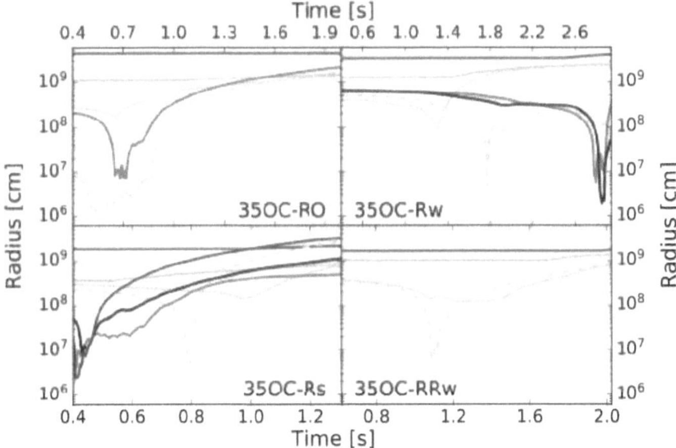

Figure 7.10.: Radius of representative tracer particles for each group. We show two tracers for the Fe group (green dashed and straight line).

that in our reference model 35OC-RO there is no similar transition from a spherical to an oblate PNS. There, the magnetic field is strong enough to cause a high aspect ratio early on. Although we also find neutron-rich matter outside the neutrinospheres, the amount is smaller and the structure of the PNS makes it less likely that this matter will enter the outflow, thus suppressing the weak r-process group in our reference model.

7.5.3. Impact of magnetic field strength

An investigation of the models 35OC-Rw, 35OC-RO, and 35OC-Rs gives insight into the impact of the magnetic field strength on the nucleosynthesis (Fig. 7.2, see also Table 7.2). When increasing the magnetic field from model 35OC-Rw to model 35OC-RO, then elements around the second r-process peak are not produced anymore. This is related to the late evolution of model 35OC-Rw, discussed above. We note, however, that this non-monotonicity, caused by the presence or absence of late neutron-rich fluid elements, only affects a small fraction of the ejecta. When these fluid elements are ignored, the distribution of the ejecta across Y_e behaves monotonically with initial field strength (left panel of Fig. 7.5).

Explosions with strong magnetic fields, like 35OC-Rs, have been suggested as a potential r-process site (e.g., Meier et al. 1976, Meyer 1994). The magnetic field produces a jet-like explosion and prompt ejection of neutron-rich material (Fig. 7.8). The formation and stability of the jets strongly depends on the magnetic field and is still under discussion (Mösta et al. 2014, Halevi and Mösta 2018). For strong magnetic fields, some neutron-rich matter is rapidly ejected by the magnetic pressure without major neutrino interactions with neutrons. Depending on these two contributions, i.e., matter ejected by magnetic or by neutrino pressure, one can produce a strong r-process when the magnetic field dominates or a weak r-process if neutrinos become important in ejecting matter. This behaviour has been investigated in detail by, e.g., Nishimura et al. (2017), Mösta et al. (2018) by artificially varying the neutrino luminosity. Our neutrino treatment is self consistent and the neutrino luminosity is not a free parameter.

In model 35OC-Rs, matter close to the neutron star and thus at small radii (purple line in Fig. 7.8) reaches high densities and temperatures (Fig. 7.4) and is promptly ejected as seen by the fast increase of the radii. Due to the fast expansion, neutrinos are unable to convert too many neutrons into protons and the electron fraction stays low, $Y_e = 0.2 - 0.3$ (left panel of Fig. 7.5). Such conditions (i.e., fast expansion and low Y_e) allow for the r-process to produce elements up to the third r-process peak (Fig. 7.4).

In addition to the r-process, model 35OC-Rs ejects matter with variable conditions leading to a large range of nucleosynthesis products (Figs. 7.5 and 7.4). The jets are very proton rich and contribute to the Fe-group producing mainly iron-group elements but also heavier by the νp-process (Fröhlich et al. 2006, Pruet et al. 2006, Wanajo 2006). Moreover, there is neutron-rich matter that is continuously ejected around the jets, producing elements up to the second r-process peak by a weak r-process (see Fig. 7.8).

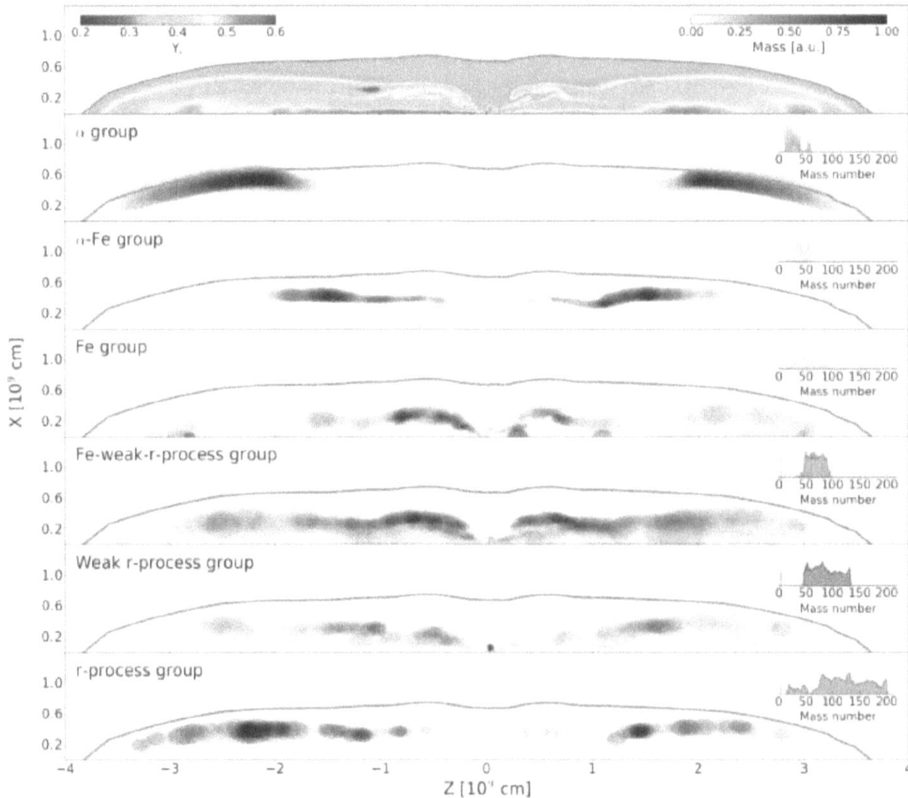

Figure 7.11.: Upper panel: Electron fraction of model 35OC-Rs at the end of the simulation with data taken from the simulation grid. The panel shows the electron fraction within the shock radius only. Outside of the shock the matter is symmetric with electron fractions of $Y_e = 0.5$. Lower panels: Mass of each individual nucleosynthesis group reconstructed from the tracer particles. Here we used the mass of each tracer together with their density to calculate the radius of a sphere representing the spatial expansion. The upper right corner of the lower panels shows the mass integrated composition of each individual group. The blue line in every panel indicates the position of the shock.

7.6. Proton-rich conditions

Besides a low electron fraction component of the ejecta, all models share the presence of a proton-rich component in the innermost part of their jet (see, e.g., the upper panel in Fig. 7.11 for the last snapshot of model 35OC-Rs or Fig. 7.9 for model 35OC-Rw). This feature is most likely connected to the relatively high fluxes of electron neutrinos, which is dictated by the sophisticated neutrino transport in combination with long-time simulations. Other studies of MR-SNe do not observe such large electron fractions, because the simulations usually stop earlier and include neutrino leakage schemes. As a consequence of these high values, all our models synthesize proton-rich nuclei (p-nuclei) that cannot be synthesized by the r-process or the s-process (see Fig. 7.3). In addition, the average mass number slightly increases with again increasing electron fraction. This is shown in Fig. 7.12 for model 35OC-RO, where we defined $\bar{A} = \sum\limits_{Z>2} X_i/Y_i$.

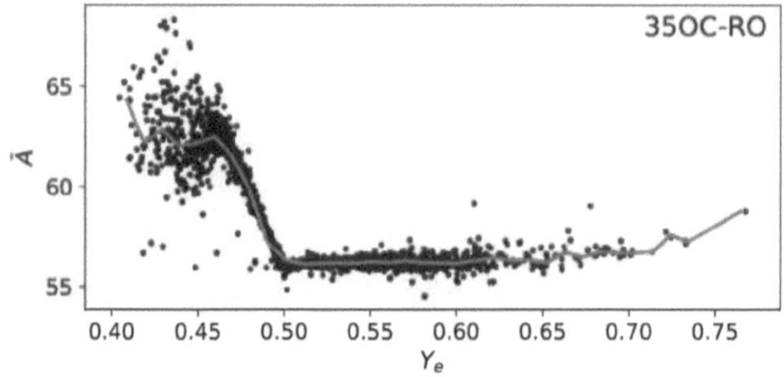

Figure 7.12.: Average mass number versus electron fraction at $5.8\,\mathrm{GK}$ of model 35OC-RO. Each point indicates a tracer particle and the blue line indicates a median value.

The most proton-rich tracer particle (i.e., $Y_e = 0.77$) of model 35OC-RO has a similar average mass number, $\bar{A}$, as tracer particles with $Y_e \sim 0.48$. The nucleosynthetic paths of these fluid elements differ, however, fundamentally. Whereas fluid elements with $Y_e \sim 0.48$ synthesize elements along stability, slightly at the neutron-rich side of the nuclear chart, the path of proton-rich fluid elements occur at the proton-rich side of the nuclear chart. Therefore, in contrast to the average mass number, the final isotopic mass fractions differ (Fig. 7.13). Of special interest is the fraction of p-nuclei such as ^{92}Mo, ^{94}Mo, ^{96}Ru, and ^{98}Ru. On the one hand, for lower electron fractions ($Y_e \approx 0.47$), (p,γ)-reactions can synthesize these nuclei. On the other hand, higher electron fractions ($Y_e \gtrsim 0.6$) synthesize p-nuclei by β^+-decays from the proton-rich side of the nuclear chart (for a detailed discussion see Arcones and Bliss 2014, Bliss et al. 2018a, Eichler et al. 2018). As indicated in Fig. 7.13, the synthesis with slightly neutron-rich conditions lead to an enhancement of Ni, caused by the magic number at $Z = 28$. In all models, the center of the jet has low densities. As a consequence, the mass of the most proton-rich ejecta is small. Considering that MR-CCSNe events may be rare, a significant contribution to the isotopic ratios of p-nuclei is questionable. Furthermore, the equation of state in the hydrodynamical simulations presented here has its boundaries at $Y_e = 0.56$. Higher values are subject to extrapolation and are therefore less certain.

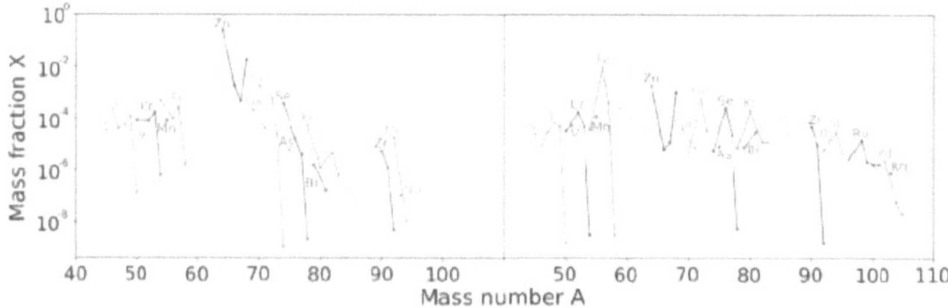

Figure 7.13.: Left panel: Mass fraction of isotopes for a fluid element with $Y_e = 0.48$. Right panel: Mass fraction of isotopes for a fluid element with $Y_e = 0.77$. Both fluid elements were taken from model 35OC-RO.

7.7. Observables

Model 35OC-Rs is the only model that synthesizes a full r-process pattern up to the third r-process peak. So far, no observation of CCSNe that produces heavy elements exist. Therefore, we can use this model in order to discuss and probe possible observables of MR-CCSNe. In addition, we will discuss possible differences to models of NSMs and suggestions to disentangle both production sites of heavy elements.

7.7.1. A comparison of ejected elements with abundance pattern of old stars

One possibility to distinguish different astrophysical hosts of the r-process is the resulting composition. Model 35OC-Rs is less neutron-rich than the dynamical ejecta of typical simulations of NSMs (e.g., Rosswog et al. 1999, Korobkin et al. 2012, Bovard et al. 2017).

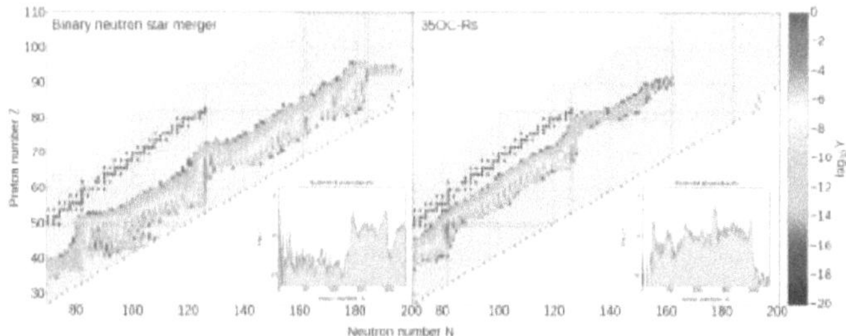

Figure 7.14.: Left panel: Nuclear chart together with abundances for a representative fluid element from the simulations of Rosswog et al. (1999) and Korobkin et al. (2012). The abundances are shown before the decay back to stability. Right panel: Same as left panel, but for a representative fluid element of the r-process group of model 35OC-Rs.

Therefore, even for the most neutron-rich fluid elements in model 35OC-Rs, the nucleosynthetic pathway proceeds closer to stability than for the dynamical ejecta of NSMs (Fig. 7.14). In general, MR-CCSNe models are currently thought to be less neutron-rich (see also, e.g., Mösta et al. 2018). The achieved neutron densities in MR-SNe as well as NSMs are, however, still a matter of debate. Nevertheless, differences in the neutron-richness may provide measurable differences in the ejecta composition of NSMs and MR-CCSNe.

Figure 7.2 shows the final mass fractions together with the solar r-process residual. Furthermore, Fig. 7.15 illustrates a comparison to the metal-poor stars CS 22892-052[4] (Sneden et al. 2003) and HD122563[5] (Honda et al. 2004) normalized to Strontium. None of our models shows a clear agreement with these observed abundances. The pattern of model 35OC-Rs seems to agree better with a Honda-star like pattern. However, due to the underlying uncertainties in the theoretical models of MR-CCSNe, nuclear physics uncertainties, as well as uncertainties in the models of stellar atmospheres (e.g., 1D or LTE assumptions) it is currently unlikely that the observations will match perfectly (for an overview of the nuclear physics uncertainties see, e.g., Eichler et al. 2015, Mendoza-Temis et al. 2015, Martin et al. 2016, Mumpower et al. 2016, Côté et al. 2018b or Horowitz et al. 2019, Arnould and Goriely 2020 for recent reviews.).

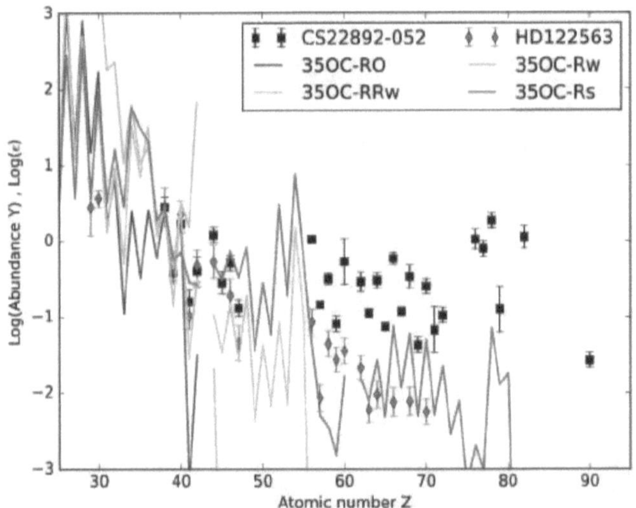

Figure 7.15.: Atomic number versus absolute abundances. All models are normalized to strontium. Green diamonds indicate the metal-poor star HD122563 (Honda et al. 2006). Black squares illustrate the abundances of CS 22892-052 (Sneden et al. 2003).

---

[4] $T_{\mathrm{eff}} = 4800 \pm 75\,\mathrm{K}$, $\log g = 1.5 \pm 0.3$, [Fe/H] $= -3.1 \pm 0.13$, and $\nu_T = 1.95 \pm 0.15\,\mathrm{km\,s^{-1}}$
[5] $T_{\mathrm{eff}} = 4570 \pm 100\,\mathrm{K}$, $\log g = 1.1 \pm 0.3$, [Fe/H] $= -2.77 \pm 0.19$, and $\nu_T = 2.2 \pm 0.5\,\mathrm{km\,s^{-1}}$

7.7.2. Absolute europium abundances compared to dwarf galaxies

An indirect observable is the amount of r-process material in UFD galaxies. The total abundance of europium Y_{Eu} and iron Y_{Fe} can be used to estimate the amount of europium in the photosphere of stars. In addition, we allow for an arbitrary amount of iron X_{Fe} that is mixed into the stars (e.g., originating in "standard" CC-SNe). To get an estimate, we also need to know the amount of matter where the ejecta is mixed in, M_{mix}. This can be taken from the initial gas mass of UFD galaxies as Reticulum II[6]. We calculate

$$[\mathrm{Eu}/\mathrm{Fe}] = \log_{10}\left(\frac{Y_{\mathrm{Eu}}/M_{\mathrm{mix}}}{X_{\mathrm{Fe}} + Y_{\mathrm{Fe}}/M_{\mathrm{mix}}}\right) - (\mathrm{Eu}_\odot - \mathrm{Fe}_\odot), \tag{7.1}$$

where we defined $Y_{\mathrm{Eu}} = M_{^{151}\mathrm{Eu}}/151 + M_{^{153}\mathrm{Eu}}/153$. Similar, we calculate a metallicity

$$[\mathrm{Fe}/\mathrm{H}] = \log_{10}\left(X_{\mathrm{Fe}} + Y_{\mathrm{Fe}}/M_{\mathrm{mix}}\right) + 12 - \mathrm{Fe}_\odot. \tag{7.2}$$

For the solar absolute abundances $\mathrm{Eu}_\odot$ and $\mathrm{Fe}_\odot$, we assume 0.52 and 7.50, respectively (Asplund et al. 2009). The resulting value for [Eu/Fe] and [Fe/H] for different M_{mix} is shown as diagonal dashed lines in Fig. 7.16.

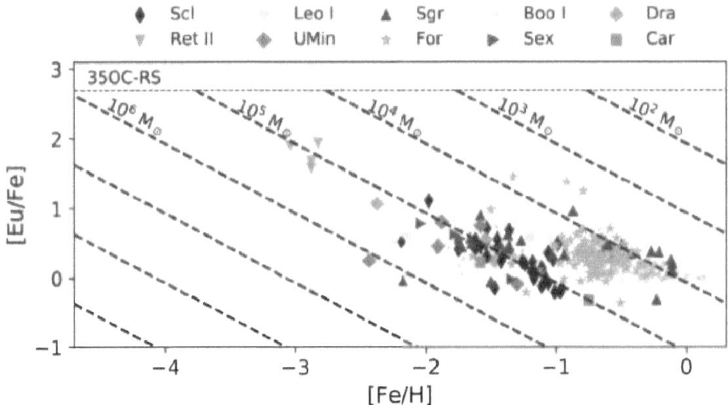

Figure 7.16.: Ejected europium and iron compared to the abundances of dwarf spheroidal galaxies. Shown are abundances from Sculptor (Scl), Leo I, Sagittarius (Sgr), Bootes I (Boo I), Draco (Dra), Reticulum II (Ret II), Ursa Minor (UMin), Fornax (For), Sextants (Sex), and Carina (Car). Grey dots are abundances in the Milky Way as in Fig. 5.12. The horizontal dashed line indicates the logarithmic europium to iron ratio relative to the sun of 35OC-Rs. Diagonal lines are ratios of equal mixing mass and arbitrary amount of iron.

The horizontal dashed line is the maximum amount of [Eu/Fe] for 35OC-Rs. Here, our abundances of chapter 2.5.2 are shown. For the example of Reticulum II, we can assume that only one r-process event contributed to the europium enrichment of the stars. Within 35OC-Rs, the amount of produced

---

[6]The initial gas mass relies on theoretical estimates.

europium is consistent with the observed values by considering an initial gas mass of Reticulum II of about $2.4 \cdot 10^5 \, M_\odot$ (Beniamini et al. 2016). Indeed, our ejected europium mass of $\sim 4 \cdot 10^{-6} \, M_\odot$ is sufficiently close to the value of $\sim 10^{-5} \, M_\odot$ that is found for Reticulum II in Beniamini et al. (2016). Another interesting aspect is the existing lower limit in metallicity. If model 35OC-Rs would be responsible for the synthesized europium in UFD, the metallicity is also constrained to a lower limit. This limit is visible for $X_{\mathrm{Fe}} = 0$ and therefore as the intersection between the diagonal and the horizontal dashed lines in Fig. 7.16. However, we want to stress that the amount of europium assumes that no additional europium is ejected from the surrounding accretion disk. The amount of ejected iron of model 35OC-Rs may also be higher with increasing simulation time. An arbitrary increase of iron is, however, already taken into account with X_{Fe} and will move the [Eu/Fe] ratio on the diagonal lines. Nevertheless, the [Eu/Fe] ratio will be lower and so will the horizontal line in Fig. 7.16. The minimum metallicity will remain the intersection of this shifted horizontal line with the unchanged diagonal lines.

7.7.3. The detection of γ- and X-rays

A direct observable are γ- and X-rays. Previous studies (Qian et al. 1998, Ripley et al. 2014, Korobkin et al. 2019) showed the potential of measuring individual emission features in CC-SNe remnants that are powered by the β-decay of heavy nuclei. Due to the velocity of the ejecta and the high and slowly decreasing opacities, the lifetime of the investigated nuclei have to be in the order of at least several decades (Ripley et al. 2014). We calculate the fluxes that result from our ejecta composition by (Qian et al. 1998):

$$F_\gamma = \frac{N_A}{4\pi d^2} \frac{M_x}{A} \frac{I_\gamma}{\bar{\tau}}, \tag{7.3}$$

where N_A is Avogadro's number, d the distance to the remnant, M_x the synthesized mass of nucleus x, I_γ the number of photons per energy emitted by the decay of the nucleus, and $\bar{\tau} = T_{1/2}/\ln(2)$ the lifetime of the investigated nucleus. For I_γ we take values obtained from the Lund/LBNL Nuclear Data Search[7], whereas we use lifetimes from the JINA Reaclib database (Cyburt et al. 2010). To calculate a spectrum, we take a representative trajectory of each above presented group and multiply the composition by the mass of the corresponding bin. Besides the strong emission lines of ^{44}Ti at 68 keV and 78 keV that have been already observed in CC-SNe as e.g., SN1987A and Cas A (e.g., Iyudin et al. 1994, Grebenev et al. 2012) also the decay of ^{137}Cs shows significant features in the spectrum (Fig. 7.17). We want to stress that our calculation is only a rough estimate as the spectrum shown in Fig. 7.17 only includes emission lines and no continuous radiation. In addition, we assumed infinite resolution of the instrument. Real instruments will have a lower maximum flux and a Gaussian distribution around the emission lines due to instrumental broadening. Nevertheless, the calculation raises hints for a direct observable, namely the emission lines of ^{137}Cs. Whether these features can be observed with current or upcoming detectors is beyond the scope of our work.

---

[7] http://nucleardata.nuclear.lu.se/toi/

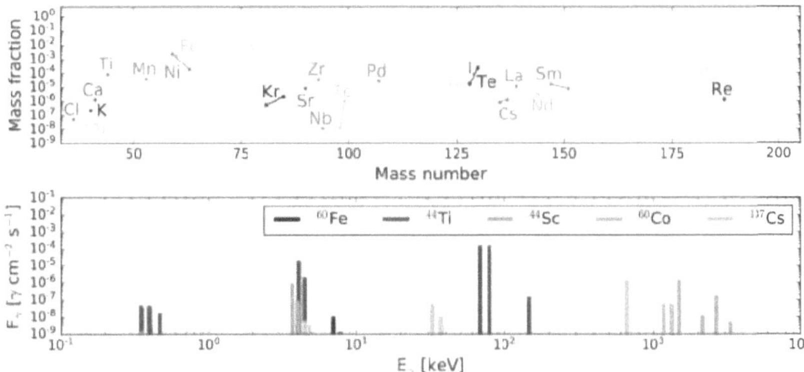

Figure 7.17.: Upper panel: Mass fraction of radioactive nuclei after 100 years. Isotopes are connected with lines and indicated in uniform colors. Lower panel: γ- and X-ray spectrum after 100 years at a distance of 10 kpc at infinite resolution.

8. Unusual abundances of neutron-capture elements in a peculiar star

Rare CC-SNe may indeed be a candidate for the production of heavy elements. These type of events may be able to explain all observational constraints. However, the detection of MR-SNe is challenging and their modeling remains uncertain. Further studies, including the investigation of more progenitors, accurate neutrino transport, 3D, and evolved for a long-time will give more accurate answers. Even when considering these rare events, the chemical signature of some individual stars is still puzzling. An example for such an odd star is [KMP2016] 10464 (in the following #10464). This star is fairly metal-poor ([Fe/H] = -1.53), located in the Galactic bulge, and shows an immense overabundance of Rb ([Rb/Fe] = 1.29 ± 0.16). It is one of only three known metal-poor, n-capture enhanced stars in the bulge (see, e.g., Johnson et al. 2013, Lucey et al. 2019) and will be analyzed in the following. This chapter was carried out within a collaboration together with A. Koch, C. J. Hansen, M. Hampel, R. J. Stancliffe, A. Karakas, and A. Arcones and is published in Koch et al. (2019).

8.1. Stellar parameters

The determination of the stellar parameters for #10464 was done in Koch et al. (2016). They employed an equivalent width analysis and enforced excitation and ionization balance to obtain the stellar parameters (Sect. 4.3). Furthermore, they assumed LTE (Sect. 4.1.1). The corresponding atmosphere model was generated using a plane-parallel, one-dimensional grid of ATLAS model atmospheres[1] and the analysis was performed with MOOG (Sneden 1973b).

Summarized, they derive T_{eff} = 5400 K, $\log g$ = 1.7, ξ_t = 2.64 km s^{-1}, and [Fe/H]= -1.53.

8.2. Abundance measurements

Most of the abundances of #10464 have already been introduced in Koch et al. (2016). Table 8.1 recapitulates the abundance measurements in this star obtained in the latter work. In addition, we were able to extract further elemental abundances which were not included in the latter study. This includes absorption lines of Li I, O I, Ga I, Ce II, Sm II, Pr II, Gd II, Tb II, Dy II, Ho II, Er II, Hf II, and Pb I (Table 8.2). The systematic errors were derived similar to Sect. 5.4.12 by varying the stellar models by one parameter about its uncertainty at a time and recalculating a new set of abundances (see also Koch et al. 2016, for more details).

---

[1] http://kurucz.harvard.edu/grids.html

Table 8.1.: Abundance ratios in the C-rich bulge star #10464 from Koch et al. (2016) and our present measurements. Table taken from Koch et al. (2019).

Element	$\log\varepsilon$	[X/Fe]	Element	$\log\varepsilon$	[X/Fe]	Element	$\log\varepsilon$	[X/Fe]	Element	$\log\varepsilon$	[X/Fe]
Li I		<0.70	Cr I	3.83	-0.28 ± 0.07	Zr II	1.35	0.30 ± 0.17	Ho II	-0.65	0.40 ± 0.20
C I	7.31	0.41 ± 0.21	Mn I	3.64	-0.26 ± 0.14	Ba II	2.00	1.35 ± 0.08	Er II	-0.01	0.60 ± 0.20
N I	7.05	0.75 ± 0.15	Fe I	5.97	-1.53 ± 0.06	La II	0.49	0.92 ± 0.15	Hf II	0.82	1.50 ± 0.20
O I	7.79	0.63 ± 0.13	Fe II	5.98	-1.52 ± 0.06	Ce II	1.30	1.24 ± 0.17	Pb I	1.72	1.50 ± 0.20
Na I	5.25	0.54 ± 0.06	Co I	3.50	0.04 ± 0.09	Pr II	0.11	0.92 ± 0.05	[C/N]	...	-0.34 ± 0.26
Mg I	6.56	0.49 ± 0.08	Ni I	4.73	0.03 ± 0.10	Nd II	1.02	1.13 ± 0.10	[N/O]	...	0.12 ± 0.20
Si I	6.56	0.58 ± 0.07	Zn I	3.34	0.31 ± 0.05	Sm II	0.33	0.90 ± 0.09	[Ba/La]	...	0.43 ± 0.17
Ca I	5.00	0.19 ± 0.10	Ga I	2.52	0.40 ± 0.20	Eu II	-0.64	0.37 ± 0.16	[Eu/La]	...	-0.55 ± 0.22
Sc II	1.62	0.00 ± 0.08	Rb I	2.28	1.29 ± 0.16	Gd II	0.09	0.55 ± 0.11	[hs/Fe]	...	1.17 ± 0.08
Ti I	3.79	0.37 ± 0.09	Sr II	2.18	0.84 ± 0.07	Tb II	-0.38	0.85 ± 0.20	[ls/Fe]	...	0.53 ± 0.07
V I	2.22	-0.18 ± 0.16	Y II	1.14	0.46 ± 0.11	Dy II	-0.05	0.38 ± 0.10	[hs/ls]	...	0.64 ± 0.11

Note: The given, total error includes a 1σ statistical and the systematic uncertainty.

Table 8.2.: Line list for elements in #10464 that were not covered in Koch et al. (2016). Table taken from Koch et al. (2019).

Element	λ [Å]	E.P. [eV]	$\log gf$	Element	λ [Å]	E.P. [eV]	$\log gf$	Element	λ [Å]	E.P. [eV]	$\log gf$
Li I	6707.80	0.00	0.17	Sm II	4815.81	0.19	-0.77	Dy II	3757.37	0.10	-0.17
O I	7771.94	9.15	0.32	Pr II	4062.80	0.42	0.33	Dy II	3944.68	0.00	0.11
O I	7774.17	9.15	0.17	Pr II	4141.22	0.55	0.38	Dy II	4103.31	0.10	-0.38
O I	7775.39	9.15	-0.05	Pr II	4143.13	0.37	0.60	Dy II	4449.70	0.00	-1.03
Ga I	4172.00	0.10	-0.31	Pr II	4179.40	0.20	0.46	Ho II	3810.71	0.00	0.19
Ce II	5274.23	1.04	0.15	Pr II	4222.95	0.06	0.23	Ho II	4045.45	0.00	-0.05
Sm II	4536.51	0.10	-1.28	Pr II	4408.81	0.00	0.05	Er II	3692.65	0.05	0.14
Sm II	4577.69	0.25	-0.65	Gd II	4130.37	0.73	-0.02	Er II	3729.52	0.00	-0.59
Sm II	4642.23	0.38	-0.46	Gd II	4251.57	0.38	-0.22	Hf II	4093.16	0.45	-1.15
Sm II	4676.90	0.04	-0.87	Tb II	4752.53	0.00	-0.55	Pb I	4057.81	1.22	-0.22

8.3. Modeling the peculiar abundance pattern

The abundances, given in Table 8.1 are shown together with a range of AGB models from the F.R.U.I.T.Y. database (Cristallo et al. 2011) for different stellar masses in Fig. 8.1. None of the models is able to reproduce all heavy-element peaks, in particular the high [Rb/Fe] ratio.

In the following, we try to disentangle the abundance contributions from various processes to the chemical enrichment of #10464. For this, we considered models of the s-process (Sect. 8.3.1), the r-process (Sect. 8.3.2), and the i-process (Sect. 8.3.3)

8.3.1. s-process yields

For the s-process, we considered the metal-poor ($Z = 0.0001$; [M/H]$=-2.2$ dex) AGB models of Lugaro et al. (2012). We took the entire, broad range of stellar masses (0.9–6 M$_\odot$) into account. Note that this set of AGB calculations also accounted for varying initial chemical compositions (e.g., in terms of varying heavy element contributions from early Galactic chemical enrichment, Kobayashi et al. 2006). Furthermore, the models include a still uncertain parameter in AGB nucleosynthesis, the size of the ^{13}C pocket (see, e.g., Buntain et al. 2017 for a detailed discussion).

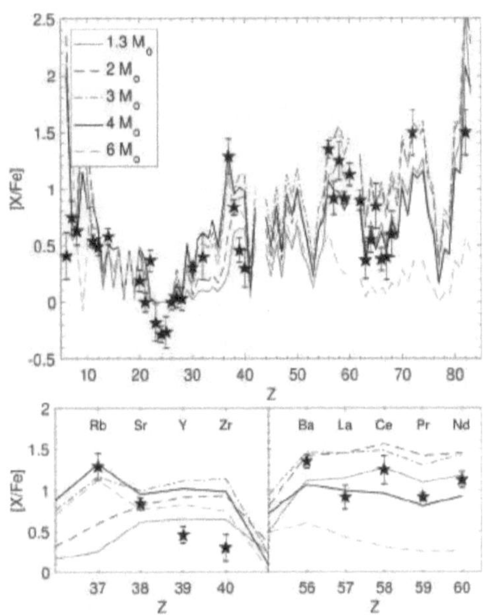

Figure 8.1.: Upper panel: Full abundance distribution of the bulge star #10464 from Koch et al. (2016) and this work. The AGB yields were taken from Cristallo et al. (2011). The bottom panels are a zoom into the regions of the first and second s-process peaks. These figures were taken from Koch et al. (2019).

In addition, tests were carried out using the more metal-rich models of Fishlock et al. (2014, Z=0.001) and Karakas and Lugaro (2016, Z=0.007 and Z=0.014). These tests resulted in considerably larger χ^2 values when fitted to the observations. Coupled with the low metallicity of the star to be described, at $[\text{Fe}/\text{H}] = -1.5 \pm 0.06\,\text{dex}$, we discard these metal-rich yields from the following considerations.

8.3.2. r-process yields

The r-process calculation was performed with the nucleosynthesis network WinNet (see Sect. 3.4) that contains ~ 6000 nuclei, including ~ 65000 reaction rates of the Jina Reaclib Database V2.0 (Cyburt et al. 2010), using the Finite-Range droplet mass model (Möller et al. 1995). In addition, we used neutron-induced fission rates given by Panov et al. (2010). Here, the r-process is calculated in the environment of the dynamical ejecta from compact neutron star mergers using temperature and density profiles from Newtonian simulations (Rosswog et al. 1999). Korobkin et al. (2012) investigated the r-process nucleosynthesis for these ejecta and found a very robust abundance pattern for heavy nuclei, caused by the low electron fractions of ~ 0.035 that leads to fission cycling (see also Eichler et al. 2015). Therefore, we choose one representative temperature and density profile to calculate the typical r-process abundances. As discussed in chapter 7, current r-process nucleosynthesis calculations are still quite uncertain. Besides uncertain astrophysical conditions, most of the nuclear reactions involved in the r-process nowadays still rely on theoretical predictions rather than experimental data. As a consequence, theoretical nucleosynthesis calculations are not able to fully reproduce the solar r-process abundances. Therefore, we also considered the abundance pattern of the metal-poor star CS 22892-052 as a reference set, assuming that its heavy elements are produced by the r-process only (Sneden et al. 2003). Even with this pattern we reach the same conclusion that the contribution of r-process to the bulge star #10464 is negligible.

8.3.3. Model of the i-process with fixed neutron exposure

With the nuclear reaction network WinNet it is not possible to fix neutron exposures. Therefore, the suite of codes "NucNet Tools" (Meyer 2012) was used under hydrostatic conditions (i.e., fixed temperature and density). The conditions are chosen as $T = 1.5 \times 10^8$ K and $\rho = 1600$ g cm^{-3}, representing the mid-point of the intershell region in a low-metallicity AGB model (see Stancliffe et al. 2011 for further details of the structure). For these low temperatures, the reaction rates do not establish an equilibrium (Sect. 3.1) and we assume the initial composition from the intershell of a low-metallicity ($Z = 0.0001$), low-mass (M = 1 M$_\odot$) AGB model after the second thermal pulse (Abate et al. 2015b, and references therein).

The nuclear reaction network considered 5442 nuclei and 45831 reactions from the JINA Reaclib V0.5 database (Cyburt et al. 2010) with further α-decay rates from Tuli (2011).

In addition, the neutron density was fixed to $N = 10^{15}$ cm^{-3} for 0.1 years. This resulted in a neutron exposure of $\tau = 495$ mbarn^{-1}. For models with differing neutron densities (i.e., $N = 10^9$ cm^{-3} to $N = 10^{15}$ cm^{-3}) the run-time was fixed to assure a constant neutron exposure (see Hampel et al. 2016, for details of the models).

This method has the advantage of ensuring a robust equilibrium-abundance pattern, but has the drawback that lead abundances (Z=82) cannot be predicted. There, the reaction flows do not yield an equilibrium, which results in implausibly high level of Pb production despite the otherwise very robust i-process pattern. This was in detail investigated in Hampel et al. (2016) and Hampel et al. (2019).

Given these complications, we explicitly remove lead from all further considerations in our statistics.

8.4. Model results

Here, we apply the setups laid out in Sect. 8.3 to test if the bulge star #10464 shows signatures indicative of s-, r-, or i-process nucleosynthesis. One of our assumptions is that the nucleosynthetic process occurred in an earlier generation of stars that pre-enriched the gas from which #10464 was formed.

We consider elements with $31 \leq Z \leq 72$ (Ga through Hf) for our fitting procedure. Lighter elements (Z≤ 30) are not significantly produced in n-capture reactions and several processes contribute to their chemical enrichment. The quality of each scenario was judged in terms of the χ^2 statistics. We calculate χ^2 within the element range of $Z_i \leq Z \leq Z_f$ by

$$\chi^2 = \sum_{Z_i}^{Z_f} \left(\log\left(\varepsilon(Z)\right) - \log\left(c \cdot Y_\mathrm{M}\right) \right)^2 / \sigma(Z)^2, \qquad (8.1)$$

where $\sigma(Z)$ is the error on the observationally derived $\log\varepsilon$ abundances (Table 8.1), and Y_{M} are the model yields from every process. The fit of one distribution is obtained by a multiplicative scaling factor c of the abundances Y_{M}, which translates into an additive scaling in logarithmic space. This is equivalent to an admixture of the individual processes with pure hydrogen.

The results for each model is summarized in Table 8.3. There, the mass of the s-process contributing AGB-star is indicated in solar masses as a subscript (e.g., "s_{m2}" for a 2 $M_\odot$ star) and the i-process is identified by the logarithm of its neutron density (e.g., i_{n9} for neutron densities of $N = 10^9\,\mathrm{cm}^{-3}$).

When assuming Gaussian errors and considering that we have N=18 elements in our fit range of $31{\leq}Z{\leq}72$, we can estimate that a statistically good fit corresponds to a χ^2 of about 40, while an excellent result should yield values of the order of 10.

A chosen set of best fits together with the abundance pattern of #10464 is shown in Fig. 8.2, Fig. 8.3, and Fig. 8.4. As indicated in Table 8.3, admixtures of the solar abundance distribution (Lodders 2003) has an adverse effect on the statistics and we do not consider this option any further.

Table 8.3.: Results for various linear combinations of nucleosynthetic processes in the fitting range of $31 \leq Z \leq 72$. This table was taken from Koch et al. (2019).

Process(es)	χ^2	Notes	Process(es)	χ^2	Notes
solar	154.72	1	$s_{m2} + i_{\mathrm{n9}}$	66.31	3,2
i_{n9}	214.25	2	$s_{m2} + i_{\mathrm{n15}}$	51.17	3,2
i_{n10}	229.76	2	$s_{m5} + i_{\mathrm{n9}}$	59.25	5,2
i_{n11}	251.94	2	$s_{m5} + i_{\mathrm{n15}}$	74.37	5,2
i_{n12}	257.21	2	solar $+ i_{\mathrm{n9}}$	92.89	1,2
i_{n13}	304.40	2	solar $+ i_{\mathrm{n10}}$	94.82	1,2
i_{n14}	463.06	2	solar $+ i_{\mathrm{n11}}$	101.47	1,2
i_{n15}	539.71	2	solar $+ i_{\mathrm{n12}}$	103.30	1,2
s_{m2}	85.90	3	solar $+ i_{\mathrm{n13}}$	109.14	1,2
s_{m3}	63.56	4	solar $+ i_{\mathrm{n14}}$	101.85	1,2
s_{m5}	274.66	5	solar $+ i_{\mathrm{n15}}$	94.44	1,2
$s_{m3} + i_{\mathrm{n9}}$	63.15	4,2	$s_{m3} + r + i_{\mathrm{n9}}$	63.00	4,6,2
$s_{m3} + i_{\mathrm{n10}}$	62.94	4,2	$s_{m3} + r + i_{\mathrm{n10}}$	62.80	4,6,2
$s_{m3} + i_{\mathrm{n11}}$	62.77	4,2	$s_{m3} + r + i_{\mathrm{n11}}$	62.72	4,6,2
$s_{m3} + i_{\mathrm{n12}}$	61.23	4,2	$s_{m3} + r + i_{\mathrm{n12}}$	61.23	4,6,2
$s_{m3} + i_{\mathrm{n13}}$	59.95	4,2	$s_{m3} + r + i_{\mathrm{n13}}$	59.92	4,6,2
$s_{m3} + i_{\mathrm{n14}}$	59.71	4,2	$s_{m3} + r + i_{\mathrm{n14}}$	59.71	4,6,2
$s_{m3} + i_{\mathrm{n15}}$	54.97	4,2	$s_{m3} + r + i_{\mathrm{n15}}$	54.57	4,6,2
$s_{m3} + r$	63.26	4,6	2-step i	50.96	7

References and model details: (1): Solar abundances from Lodders (2003); (2) i-process abundances from Hampel et al. (2016), using constant temperatures of $T = 0.15\,\mathrm{GK}$ and constant densities of $\rho = 1600\,\mathrm{g\,cm}^{-3}$. The respective neutron densities are indicated (as logN [cm^{-3}]) by the subscript; (3): AGB yields for $M_{\mathrm{init}} = 2\,M_\odot$, $Z = 0.0001$ (Lugaro et al. 2012); (4): AGB yields for $M_{\mathrm{init}} = 3\,M_\odot$, $Z = 0.0001$ (Lugaro et al. 2012); (5): AGB yields for $M_{\mathrm{init}} = 5\,M_\odot$, $Z = 0.0001$ (Lugaro et al. 2012); (6): r-process from dynamical ejecta of binary neutron star merger (Korobkin et al. 2012); (7): i-process with two ingestion episodes of τ=0.30 and 0.96 mbarn^{-1}.

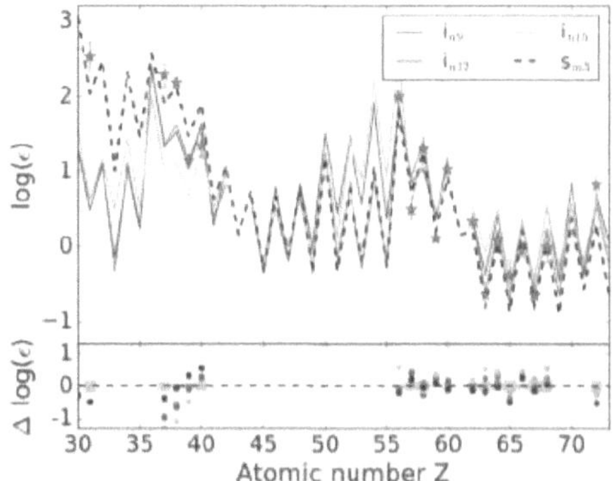

Figure 8.2.: Heavy element pattern in #10464 in comparison with i-process model calculations for various neutron densities reaching from $N = 10^9\,\mathrm{cm}^{-3}$ to $N = 10^{15}\,\mathrm{cm}^{-3}$. Here, the s-process curve is for the best-fit 3 $M_\odot$ AGB composition of Lugaro et al. (2012). The top panel shows absolute abundances, the lower panel the resulting residuals of the individual fits and the errorbars of the abundances as squares. Pb (Z=82) was excluded from our statistical tests and is thus not shown in this and the following figures (Koch et al. 2019).

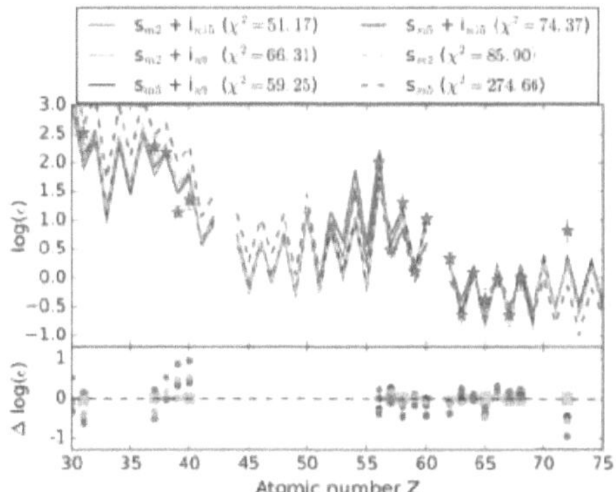

Figure 8.3.: Same as Fig. 8.2, but for various mixtures of s-process with i-process models. Shown are the curves for the overall, best-fit version ($s_{m2} + i_{n15}$; blue), a 2 $M_\odot$ AGB component plus a lower neutron density (red), and an s-process from a high-mass AGB model plus the two extreme neutron density i-processes (black and magenta). Shown as dashed lines are two s-process models from different AGB masses (Koch et al. 2019).

8.5. s- versus i-process

A pure dilution with hydrogen for the best-fit s-process from the AGB models of Lugaro et al. (2012) is shown in Fig. 8.2. In addition, we show i-process models with neutron densities of $N = 10^9$ cm^{-3} up to 10^{15} cm^{-3} (Hampel et al. 2016).

Here, our fitting emphasized that the pure s-process pattern of the Lugaro et al. (2012) yields provides a good agreement with the observed data (see also Table 8.3). The progenitor of the best fitting model has an initial mass (M_{init}) of $3\,\mathrm{M}_\odot$ and $M_{\mathrm{evol}} = 2.51\,\mathrm{M}_\odot$ after its evolution[2]. This model is characterized by core and envelope masses of $\mathrm{M_{Core}} = 0.81\,\mathrm{M}_\odot$ and $\mathrm{M_{Env.}} = 1.70\,\mathrm{M}_\odot$, respectively, and implies a fairly massive white dwarf companion.

In comparison to a pure s-process enrichment, an i-process enrichment provides a larger χ^2, which was smallest for a neutron density of $N = 10^9$ cm^{-3}. Overall, the χ^2 values for these scenarios are of the order of hundreds. Therefore, we consider the fits as bad and statistically insignificant.

We note that also the s-process pattern implies an overproduction of lead as described for the i-process in Sect. 8.3.3. This is however not reflected in our fitting procedure, because we explicitly excluded lead.

As suggested by Denissenkov et al. (2017), rapidly accreting white dwarfs may account for about a third of the intermediate n-capture elements ($31 \leq Z \leq 42$) for stars with higher metallicities ([Fe/H]> -1 dex). However, none of the abundances of the bulge star #10464 reach the predicted over-enhancements in that is predicted in this scenario (cf. their Fig. 4).

Inferred by the high [Rb/Zr] ratio, Koch et al. (2016) suggested that the AGB progenitor was likely of intermediate mass ($\sim 4\,\mathrm{M}_\odot$). A detailed match could not be achieved for the remaining abundance pattern (of 10 elements with Z$\geq$30). The models also fail to describe the very high [Rb/Fe] abundance of this star. The trend of strongly decreasing [ls/Fe] ratios when moving from Rb to Zr is even more puzzling (Fig. 8.1). These features are not reproduced in any of our models and pose a challenge to nucleosynthetic calculations.

Such a decrease in the ratios from Sr through Zr is seen in models of fast rotating massive stars ($\sim 25\,\mathrm{M}_\odot$, e.g., Frischknecht et al. 2012). However, those stars do not produce a high [Rb/Fe] ratio. High [Rb/Fe] ratios are predominantly produced in AGB stars, but a superposition of such enrichment with the more regular intermediate-mass pollution (van Raai et al. 2012) seems unlikely.

We compared the Rb abundances with a $\sim 6\,\mathrm{M}_\odot$ AGB-model of Pérez-Mesa et al. (2017) and were able to reproduce the abundance of Rb. However, matching Rb for these models stands in conflict with too low [Rb/Zr] ratios in comparison to our observation of #10464. Also other high-mass AGB models suffer from this overproduction of Zr ("s_{m5}" in Fig. 8.3 and Table 8.3).

A neutron burst that is short enough to not establish a typical i-process equilibrium-abundance pattern only synthesizes elements up to the ls peak. A scenario with a neutron density of $N = 10^9$ cm^{-3} and an exposure of $\tau = 0.3$ mbarn^{-1} is able to reproduce the observed characteristics of the Rb peak. This however excludes the production of heavier elements including the hs-peak elements and Pb as further n-captures would destroy the pattern. Therefore, the i-process is not able to produce Rb in combination with heavier elements in one single event.

---

[2]Fitting the entire suite of F.R.U.I.T.Y. models yielded a lower AGB mass of 1.5 M$_\odot$, albeit at a poorer match in metallicity so that we did not pursue this comparison any further.

None of the models is able to reproduce the shape of the heavy-s peak (e.g., the observed [Ba/La] ratio), which renders a pure s-process origin unlikely. This is supported when drawing #10464 into the [Ba/La] vs. [Eu/La] plane (Fig. 6 in Mishenina et al. 2015). In this plane, our star is located at the lowest boundary of open cluster and Galactic disk stars [Eu/La] values. A general feature of i-process models is that they succeed in reproducing a higher [Ba/La] compared to the s- or r-process (e.g., Hampel et al. 2016). For #10464, the [Ba/La] ratio is too low for a substantial i-process contribution, characterized by $N = 10^{15}$ cm^{-3}. In addition the hs-peak of #10464 is remarkable, since [Ba/La]$\gg$[Ba/Ce]. An increased neutron density shapes the hs-peak predominantly through contributions of additional Ba resulting from the decay of radioactive ^{135}I. This causes troubles explaining both the Ba- and Ce-to-La ratio being 0.3 dex higher than the solar ratios.

8.6. Multiple enrichment sites

So far, we only considered one nucleosynthetic event to describe the abundance pattern of #10464. In the following, we will focus on the possibility of several enrichment sites. For this, we consider that up to three processes contributed to the chemical enrichment of the star and pre-enriched the material of which the star was formed.

Therefore, we adopt a linear superposition of N individual nucleosynthetic processes, j, following the formalism of Hansen et al. (2014b):

$$Y_{\mathrm{M}}(Z) = \sum_{j=1}^{N} c_j Y_j(Z), \qquad \text{with } c_j \geq 0 \tag{8.2}$$

where $Y_j(Z)$ denotes the absolute abundances and c_j are the weights assigned to each of the contributing processes (s,r,i), respectively. Since c_j is an arbitrary scaling factor, the actual values of these weights have no physical meaning. Implementing these scaling factors without additional constraints will cause an additive freedom. This freedom can be interpreted as a mixture with pure hydrogen. Therefore other mixing techniques as shown in, e.g., Hampel et al. (2016), introduce the constrain of $\sum c_j = 1$. Here, we want to stress that the choice of the mixing techniques does not affect the conclusion of this work.

We again minimize Eq. (8.1), but substitute Y_{M} with Eq. (8.2). The results for $Z_i = 31$ and $Z_f = 72$ are given in Table 8.3. As in Sect. 8.4, we used the models described in Sect. 8.3. In addition, we considered models from the F.R.U.I.T.Y database. In total, we tested more than 10000 different model combinations. Exemplary, Fig. 8.4 shows the best-fit of linear combinations of the s-, i-, and r-process.

Generally, a linear admixture of other processes to the s-process prescription improve the fits slightly. We obtained the lowest χ^2 for a combination of s-, and i-process. For these models, we assumed a metal-poor AGB star and the i-process that act as a mere perturbation on top of the AGB yields. For all cases, the "best" match was obtained by invoking the i-process with the highest neutron density (i_{n15}).

On the other hand, the most neutron rich scenario (the r-process) leads again to no significant improvement of the χ^2 (labeled "s+r+i"), regardless if we chose a theoretical calculation or abundances from an r-process enhanced star (Sneden et al. 2003). Moreover as indicated by Fig. 8.4, the best fit was achieved without any fraction of the r-process. Note that none of our combinations is able to reproduce the high amount of Rb and Sr.

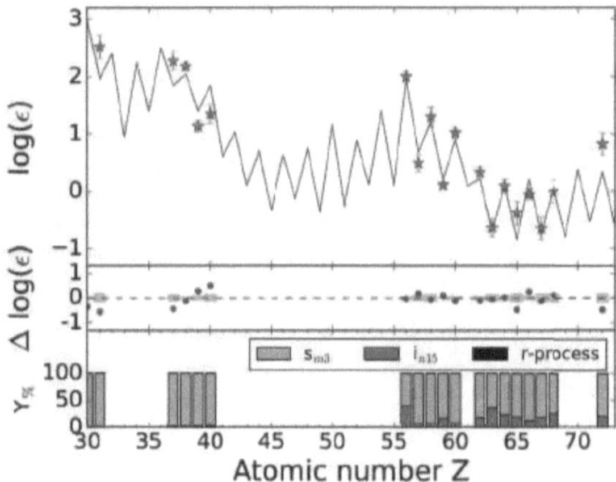

Figure 8.4.: Same as Fig. 8.2, but for a linear combination of all three nucleosynthetic channels (s+r+i). The bottom panel indicates the relative contributions from each process. Note that no r-process component is required (Koch et al. 2019).

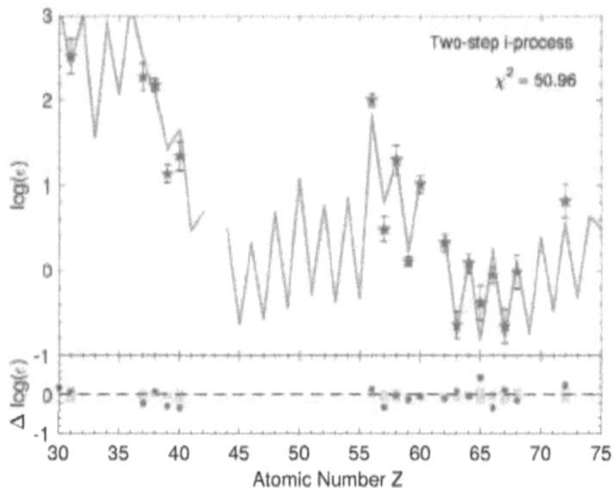

Figure 8.5.: Same as Fig. 8.2, but for a two-step i-process with two separate ingestion events of $\tau = 0.30$ and 0.96 mbarn^{-1}. Such a superposition is better able to reproduce both the light and heavy s-peaks (Koch et al. 2019).

We therefore considered a contribution of two distinct proton-ingestion events in the same donor, each of different strength. Our current understanding of the site(s) of the i-process does not allow us to make firm constraints on the exact number of successive proton-ingestion events and it has been shown that, for example, super-AGB stars could host multiple such events (Jones et al. 2016). A shorter neutron bursts with $\tau = 0.30$ mbarn^{-1} can reproduce the light s (ls) peak and a separate event with $\tau = 0.96$ mbarn^{-1} gives the best fit to the observed abundances of elements with $Z > 50$ (Fig. 8.5).

Separate bursts are required as the peak abundance moved to higher Z as the exposure increases. This builds up the ls-peak, but then moves on to the heavy s (hs) peak. For high enough exposure, lead is built up. This happens in a similar fashion as elements are produced in the s-process.

A combination of these two individual events thereby leads to the overall, best (in a χ^2-sense) explanation of the peculiar abundance pattern of #10464 (Fig. 8.5), although the high complexity of this scenario renders it, statistically, equally (im-)probable as a $3\,M_\odot$ AGB pollution plus single i-process contributions. At respective χ^2 values on the order of 50 vs. 60, the differences are marginal.

Typically, in abundance fitting, excellent χ^2 statistics as low as $\sim$10 can be reached (see also Hansen et al. 2014b, Abate et al. 2015a). Within our analysis, we did not achieve such low values (Table 8.3), which indicates that the composition of this star is not fully understood, yet, and cannot be explained satisfactorily with any of the processes (or combinations thereof) considered.

8.7. The impact of self-pollution

The stellar parameters (Sect. 8.1) of T_eff=5400 K and $\log g$=1.7 that are derived by Koch et al. (2016) place this star on the horizontal branch. The evolutionary tracks indicate a mass of $\sim 0.55\,M_\odot$ (e.g., Cassisi et al. 2004, Hansen et al. 2011). As Koch et al. (2016) based the surface gravities of the star on ionization equilibrium (see Sect. 4.3.2), we are able to determine the distance to #10464. With this information, we further estimate that the star has a luminosity of $\sim 220\,L_\odot$.

This indicates that the star is in a fairly evolved state and has undergone deep evolutionary mixing toward the tip of the red giant branch. As a consequence, it will have altered its surface composition. The evolutionary calculations of Placco et al. (2014) suggest a upward correction in [C/Fe] on the order of 0.2 dex, bringing the carbon ratio of #10464 to $\sim$0.6 dex (see also Henkel et al. 2018).

In addition to the moderate carbon enhancement, we found a strong enhancement in nitrogen (Table 8.1) of $[C/N] = -0.34 \pm 0.26$ dex. When we account for the aforementioned correction for stellar evolution this ratio even lowers to $[C/N] =\sim -0.54$ dex, placing it close to the boundary of -0.6 dex that separates evolved, mixed stars from objects that are unaffected by mixing (Spite et al. 2005, Hansen et al. 2016).

We refer the reader to Koch et al. (2019) for a more detailed discussion about the possibility of self-pollution. Here, we only want to briefly mention that a strong level of self-pollution would indicate that the presently observed surface abundance has been significantly altered from its initial composition. Therefore, the abundance peculiarities may not be indicative for external polluters only.

9. Summary and Outlook

In this book, we investigated the origin of r-process elements. In our study, we combined stellar observations, specifically stellar abundances with the nucleosynthesis of hydrodynamical models with the goal to better understand the r-process. We derived homogenized stellar abundances of dSph galaxies to study trends in their chemical enrichment. For this purpose, we conducted the largest homogenized and high-resolution study of dSph galaxies to date (Reichert et al. 2020). By having the stellar parameters of our sample stars homogeneously analyzed, we use this information to verify an empirical relationship for the determination of the stellar parameters of CEMP-stars (Singh et al. 2020). We assessed the puzzling heavy elements pattern of a peculiar metal-poor star by a superposition of several processes (Koch et al. 2019). We also analyzed stars of the Milky Way and summarized properties of the r-process host event (Côté et al. 2019a). Furthermore, we analyzed the nucleosynthesis of a possible r-process host candidate, MR-SNe. We achieved the first nucleosynthesis analysis of a long-time simulation with sophisticated neutrino transport and found elements up to the third r-process peak formed for the strongest magnetized model (Reichert et al. in preparation).

We investigated the heavy elements pattern of a metal-poor Galactic bulge star ($[Fe/H] = -1.53$). The high amount of rubidium in the composition of the star was puzzling. Therefore, we tried to explain the composition of this star with a combination of processes, i.e., s-process, r-process, and i-process. In total we tested more than 10000 combinations and the best fit was obtained by mixing the pattern of two distinct i-processes that may originate from two hydrogen ingestions into the intershell of the star. This scenario is, however, rather unlikely and we did not achieve a satisfactory agreement of other scenarios. We therefore came to the conclusion that the composition may be a result of a strong self pollution of the atmosphere layer with deeper layers of the star. As a consequence of pollution, the abundance peculiarities may not be explained by external sources only. To get a satisfying explanation for the heavy element pattern, also internal sources have to be taken into account. This was, however, beyond the scope of our study and could explain the poor agreement with our models of external pollution.

We collected reduced spectra from the publicly available ESO and KOA archives and found 380 member stars of 13 dSph galaxies. In a first step, we derived the stellar parameters (i.e., effective temperature, surface gravity, metallicity, and microturbulence) for all stars. We used as few, self-consistent, methods as possible to maximize the homogeneity of our sample. By using only two different methods to derive the stellar parameters, we tried to avoid unphysical abundance trends that could be caused by using numerous different methods with different abundance dependencies. Using the stellar parameters as an input, we derived the abundances for 12 elements (i.e., magnesium, scandium, titanium, chromium, manganese, iron, nickel, zinc, strontium, yttrium, barium, and europium). Afterwards, we investigated trends within these abundances. As an example, we explored the behavior of α-elements as a function of metallicity ($[Fe/H]$) or time. On one hand, iron serves as indicator of when type Ia SNe began to contribute to the chemical enrichment of the element. On the other hand, hydrostatic α-element abundances (e.g., Mg) are tightly correlated with the stellar mass of the CC-SN progenitor. The metallicity at which a kink ("knee") is visible in the $[\alpha/Fe]$ trend is hence a testament of the type Ia SNe time scale in the galaxy but at the same time the ratio at early times is a tracer of the progenitor masses of the CC-SNe population. The ensemble of the stellar α-element content is in addition correlated with the stellar mass of the galaxy, and we can compare the dSph galaxies and their total masses by simply exploring their α-element trends. This revealed a regularity that seems approximately to be valid even for some UFD galaxies such as Reticulum II. This shows how important the stellar mass of the galaxy and thus the environment of the stars is for the enrichment of elements. Similar to the α-elements, the trend of barium versus europium shows a clear kink, which is most likely connected with a contribution from the s-process. The trend of the r-process component of barium versus magnesium revealed a fairly flat slope. This can be a hint towards rare CC-SNe as a production channel of barium at low metallicity (Reichert et al. 2020). This is an important result, as it can constrain the input of GCE models such as the rate of the host event, or the delay time. Moreover, we could not find any remarkable differences of the trend with respect to the stellar mass of the galaxy and suggest that enrichment of r-process elements should occur in a similar fashion in a large range of stellar masses ($2.9 \cdot 10^5 \mathrm{M}_\odot$ - $2.1 \cdot 10^7 \mathrm{M}_\odot$).

Most of the investigated dSph galaxies show a decreasing trend in [Eu/Fe] versus [Fe/H] for intermediate to high metallicities. This trend has also been observed and investigated for stars in the Milky Way. From a Galactic chemical evolution perspective, this trend is difficult to explain. Regardless of the complexity of the simulation, a decreasing trend can only be achieved when changing the DTD function of NSMs (Côté et al. 2019a). The DTD function of NSMs is thought to be connected with the DTD function of short GRBs and is therefore an observable quantity. We discussed the pros and cons of different DTDs under the assumption that NSMs are the only source of r-process elements. We did this by discussing discrepancies between our assumption of the DTD function, observations, including results from population synthesis, Galactical chemical evolution, and nuclear astrophysics. The ability of these different scenarios to explain observational evidences was finally condensed and summarized in table 6.1. We came to the conclusion that most arguments point towards a DTD of t^{-1}. This stands, however, in contradiction with the observed decreasing trend of [Eu/Fe] versus [Fe/H]. A possibility to fulfill all constraints and be able to model the decrease in parallel, is to consider a second source of r-process elements. Under our assumptions, this source is more pronounced in the early universe and fade out until today. This unknown production site may have produced 60 % and 40 % of the Eu during the first 100 Myr and 1 Gyr of Galactic evolution, respectively. The idea of an additional source is not new, however, from our estimate this source has to strictly fade away as a function of time and metallicity.

A candidate that can fulfill many of the gathered criteria are MR-SNe. Due to lower mass loss and the therefore inferred higher rotation rates in the early Universe, they are thought to have occurred more frequently at earlier times (see Sect. 2.3.1). Therefore, they are a promising candidate for the unknown additional source of r-process elements. Within a comprehensive nucleosynthesis study we tackled the question if they are able to produce r-process elements from a theoretical point of view. We analyzed the ejecta composition of four hydrodynamical simulations of these events. The simulations are the first long-time simulations (up to $\sim 2.5\,\mathrm{s}$) with sophisticated neutrino treatment (presented in Obergaulinger and Aloy 2020). The progenitor for this study was kept constant, but the magnetic field strength and rotation rates were varied. In this way, it was possible to study the dependence of rotation and magnetic field strength on the composition of the ejecta (Reichert et al. in preparation). The ejecta was traced by 6570, 7272, 17446, and 2218 Lagrangian tracer particles. In total, our calculation involved 6545 nuclei and ~ 75000 reactions.

Our reference model (35OC-RO), with magnetic field strength and rotation rates taken from the stellar evolution calculation, synthesized elements approximately up to zirconium ($Z = 40$). Higher rotation rates (model 35OC-RRw) lead to less neutron-rich conditions and on average less heavy elements. However, this model has a high molybdenum enrichment ($Z = 42$) as well. Compared to the reference model, model 35OC-Rw has a reduced magnetic field strength (10^{10} G) and it synthesizes elements up to the second r-process peak. Our findings showed that a stronger magnetic field (10^{12} G) provided the necessary conditions for a full r-process up to the third peak in model 35OC-Rs.

We investigated the origin and the conditions for the synthesis of elements in the individual models. We grouped the final composition of the individual Lagrangian tracer particles per model into different bins. Even though based on the final composition, these bins separated the neutron-richness of the fluid element for the hot tracers (i.e., $T > 5.8\,\mathrm{GK}$) and the density and temperature for colder tracers. Our results demonstrate the extraordinary importance of the neutron-richness and thus the neutrino treatment for the synthesis of heavy elements. The bins were universal across the different models, but not all models host all groups. Proton-rich conditions are located close to the center of the jet, whereas more neutron-rich conditions are ejected earlier and in a cocoon around the jets. The neutron-rich ejecta of model 35OC-Rw is an exception, where the PNS changes its shape from prolate to oblate due to late angular momentum transportation to the outer layers, causing a late neutron-rich outflow. We could investigate this feature only, because the simulation was calculated for a long time ($\sim 2.5\,\mathrm{s}$). Long time simulations are important to catch all relevant features and fingerprints in the ejecta of those events.

We also investigated the properties of all Lagrangian tracer particles that were able to synthesis r-process elements. This group was exclusively hosted by the model with the strongest magnetization, 35OC-Rs. The amount of r-process material is fully determined by the first moments after core bounce as it needs fairly neutron-rich conditions to occur ($Y_e \lesssim 0.3$). However, even this model did not contain material with $Y_e < 0.2$ (at $5.8\,\mathrm{GK}$) and none of the tracer particles synthesized elements up to the fission region. Therefore, the fission treatment did not play an important role in our calculations and the mass fractions of actinides are low ($X < 10^{-8}$).

In a next step, we tried to identify observable features of model 35OC-Rs. The final nucleosynthetic pattern shows a decreasing heavy element trend that is similar to a Honda type star, e.g., HD122563 (Honda et al. 2006). The total amount of synthesized europium ($\sim 4 \cdot 10^{-4}\mathrm{M}_\odot$) as well as the minimum inferred metallicity is in agreement with our previous measurements of stars in the UFD galaxy Reticulum II. We note, however, that even at the end of the simulation iron is still synthesized and the calculated iron abundance is only a lower limit of the total ejected iron. Nevertheless, it seems unlikely that model 35OC-Rs synthesizes a multiple (at least one order of magnitude) of the already synthesized iron to cause any discrepancies with the lowest metallicities in Reticulum II. Other direct observables are the γ- and X-ray signatures. Inspired by the work of Qian et al. (1998), Ripley et al. (2014), and Korobkin et al. (2019), we investigated the possible γ- and X-ray signatures in the remnant of model 35OC-Rs. We choose a representative fluid element and investigated the abundance of unstable nuclei. After 100 years, there are still radioactive nuclei present. In addition to the emission lines of ^{44}Ti that have been already observed in CC-SNe remnants, we identify also a feature of the decay of the heavy ^{137}Cs in the spectrum. This feature may be observable with future detectors, if the event occurred sufficiently close.

To understand all aspects of the synthesis of r-process elements, it is important to look at both theoretical and observational work from a broad perspective. Within this book, we have made progress in the homogenization of stellar abundances in dSph galaxies, which contributes to a better understanding of these environments and ultimately provides insights into the enrichment of heavy elements in different galaxies. Based on state-of-the-art SN simulations with sophisticated neutrino transport and following the ejecta for more than one second after the explosion, we additionally made an important step in the calculation of the nucleosynthesis of MR-SNe. Our studies were conducted at the edge of what is possible today and contributed to a better understanding of the nature of the r-process from both an observational and theoretical perspective.

The increasing amount of computational resources and computing power will allow for more realistic hydrodynamical models, involving magnetic fields, 3D effects, and higher resolution. In addition, it will be possible to develop models that involve a sophisticated neutrino transport and longer simulation times, whose importance and further necessity we have shown in this book. On one hand, experiments on (future) collider facilities (e.g., FAIR, FRIB, and RIKEN) will provide measurements for many involved reactions, which can significantly improve the precision and accuracy of nucleosynthesis calculations. Both combined, better hydrodynamical models and more precise nucleosynthesis calculations can therefore offer a better starting point for the search for possible observables. On the other hand, additional measurements of atomic properties (e.g., the oscillator strength) and future telescopes (e.g., ELT, GMT, JWST, ATHENA) will help to better constrain the derived abundances in stars. Furthermore, gravitational wave detectors will constrain the properties of NSMs and therefore help to understand if there is room for an additional r-process production site. But it is not only better instruments that play an important role in the further development of astronomy, computing power is just as important. Calculations that include all three dimensions and calculations that provide grids for correction terms to account for LTE effects will further help to eliminate unphysical abundance trends, which we also tried to reduce within this book. Depending on the complexity of the atom, this can be quite computational demanding, but the foundation has already been laid. Combining all future developments will reveal the nature of the r-process in greater detail, including its astrophysical host(s) and provide a deep understanding of the origin of elements.